Morphometry of the Human Lung

by

Ewald R. Weibel, M. D.

Research Associate, Department of Cytology, The Rockefeller Institute, New York, N. Y.
Formerly Research Associate, Department of Medicine,
College of Physicians and Surgeons, Columbia University,
and The Cardiopulmonary Laboratory
of The Chest Service, Bellevue Hospital, New York, N. Y.

With a Foreword

by

André F. Cournand, M. D.

and

Dickinson W. Richards, M. D.

College of Physicians and Surgeons, Columbia University, New York, N. Y.

With 109 Figures

Springer-Verlag Berlin Heidelberg GmbH

1963

© by Springer-Verlag Berlin Heidelberg 1963

Originally published by Springer Verlag OHG Berlin Gottigen Heidelberg in 1963.
Softcover reprint of the hardcover 1st edition 1963
Library of Congress Catalog Card Number 63—191 91

ISBN 978-3-642-87555-7   ISBN 978-3-642-87553-3 (eBook)
DOI 10.1007/978-3-642-87553-3

C'est à vous, cher maître,
Que je voue cet oeuvre.
Vos pensées en étaient les racines –
Par votre génie lui pousseront des fleurs.
J'espère que le tronc et les tiges vous plaisent.

TO MY MASTER AND FRIEND

DOMINGO M. GOMEZ

Foreword

The work presented in this monograph marks a new era, we believe, both in the development of quantitative anatomy of the lung, and in the correlation of anatomy with physiology.

For many years, physiologists interested in the overall functioning of the lung have felt a need for better quantitative descriptions of pulmonary anatomy. As physiologists, we know a good deal about the forces operating to produce pulmonary ventilation, and the quantities that define this function in rest and exercise; and the same for effective distribution of air within the lung — "alveolar" ventilation —, and for the exchange of respiratory gases between air and blood. There have been no correspondingly precise quantitative measurements of the pulmonary structures that serve theese functions.

The great advances in the study of pulmonary anatomy in the past decade have been chiefly in the realm of "fine structure". This has tended to bring together anatomy and biochemistry or physical chemistry, rather than anatomy and physiology. This conjunction has aided, for example, the conception of diffusion as a physicochemical process, but not that of diffusion as a metabolic bodily function.

It was, therefore, a remarkably fortunate circumstance which brought together in our laboratory, about three years ago, Professor DOMINGO GOMEZ and Dr. EWALD R. WEIBEL: Professor GOMEZ a mathematician and biophysicist of distinction and long experience; Dr. WEIBEL a young anatomist trained under Professor GIAN TÖNDURY in Zürich, and with additional research experience with Professor AVERILL LIEBOW at Yale.

As the pages of this monograph will set forth, an entirely new methodology for quantitative description of a complex organ has been developed. To Dr. WEIBEL's own industry and imaginative effort are due new methods of preparation of tissues, new methods of random sampling, and adaptation of methodologies from other fields of science. His collaboration with Professor GOMEZ has led to original mathematical formulations for the translation of the data into the desired geometric forms and a new technique of counting. — Briefly summarized, this work has resulted in reliable information as to the size and number of the basic elements of pulmonary anatomy. More precision as to their number, dimensions, mode of distribution and geometric forms, should eventually facilitate mathematical and physical considerations regarding the function of the lungs. Application of these quantitative methods to the study of pathologic specimens will also provide a most timely renovation of morphologic pathology.

This new era which Drs. WEIBEL and GOMEZ have initiated in the precise quantitative description of the anatomy of the lung, may well be equally applicable to the study of other organs, and may thus eventually establish a broad community of interest between anatomy, pathology, and physiology.

It has been a privilege to have Dr. WEIBEL as one of our research associates for two years, and most gratifying to have had this important work originated and carried out in the Cardio-Pulmonary Laboratory of the Columbia University Division at Bellevue Hospital.

ANDRÉ COURNAND
DICKINSON W. RICHARDS

New York, March 1963

Preface

In discussing with physiologists possible relationships between structure and function of an organ such as the human lung, a morphologist often feels that he speaks a language different in accent and sometimes in vocabulary from that of his colleagues of the other discipline. What appears most interesting — and therefore most important — to him, may be only of secondary significance to the physiologist. In part this may be due to the fact that a morphologist is inclined to marvel at the diversity of structures used to build an integrated organism, while the physiologist will tend to seek general and simple laws underlying complex events, even if this is possible only in the sense of a first and crude approximation.

Intrigued by this apparent incongruity of the goals of pulmonary morphologists and physiologists in particular, I welcomed the prospect of collaborating as a morphologist with a group of outstanding physiologists, and I am most grateful to Dr. ANDRÉ F. COURNAND and Dr. DICKINSON W. RICHARDS for arranging this unusual opportunity for me. While I worked in the Cardiopulmonary Laboratory of Columbia University at Bellevue Hospital they generously supported and stimulated the development of a morphologic study on the lung, whose original goals were necessarily quite obscure. In the course of numerous discussions with their group a long-range program finally evolved: it asked for the search of general principles underlying the structural organization and determining the functional adequacy of the human lung. This appeared to be best achieved through a systematic quantitative analysis of pulmonary structures.

At this point I wish to express my very special gratitude to Dr. DOMINGO M. GOMEZ, to whom this book is dedicated. He actually guided the development of this work onto and along the path it eventually took by first asking the right questions, and then lending a generous helping mind when their answer was to be sought. His role in this work is ubiquitous and cannot be overappreciated. I hope, moreover, that his influence will be a lasting one.

I also wish to acknowledge my sincere gratitude to Dr. GEORGE E. PALADE for allowing me to work and learn in his Department of Cytology at The Rockefeller Institute during this last year and for generously supporting and stimulating the continuation of these studies at the cellular and subcellular levels. I would like to thank Mr. BRUCE W. KNIGHT, Affiliate at The Rockefeller Institute, for his cooperation in the development of necessary measuring principles during this period.

I am thankful to the following granting institutions for supporting this work with generous funds: to the National Institutes of Health for awarding me a Research Fellowship (FFG-47); to the Health Research Council of the City of New York for granting me two Investigatorships (I-126 and I-188); to the New York Heart Association for a Research Grant. Further funds were provided through Training Grant H-2001 of the National Heart Institute, National Institutes of Health, to the Cardiopulmonary Laboratory at Bellevue Hospital.

Finally I wish to express my sincere appreciation to those who enhanced the progress of this work in many ways: to Mrs. BARBARA B. FRANK for her enthusiastic assistance; to Dr. AVERILL A. LIEBOW, Dr. THANE ASCH and Dr. ISRAEL STEINBERG for allowing me to use some of their preparations; to Miss JEAN SEIBERT, Miss CHRISTINE ROSNER and to my wife for helping me with the preparation of the manuscript; to Dr. CHARLES E. LYGHT and Mrs. LOUELLA NIGH for revising the text; and to all my colleagues for numerous suggestions and criticisms.

New York, June 1962   E. R. W.

Table of Contents

Chapter I

Introduction

The present monograph deals with a systematic approach to a quantitative morphologic analysis of the architecture of the human lung. It discusses general problems of methodology and reports the results obtained on limited but carefully chosen material, which allowed, in particular, the derivation of reliable dimensional models of this organ. These studies were started a few years ago, while the author worked in the laboratory of Dr. ANDRÉ F. COURNAND and Dr. DICKINSON W. RICHARDS in close collaboration with Dr. DOMINGO M. GOMEZ on questions of correlation of structure and function in the lung.

The physico-mathematical approach to physiology pursued by Dr. GOMEZ requires accurate quantitative information on the dimensions of human pulmonary structure before an attempt can be made to interpret experimental physiologic data in terms of some theoretically derived physical laws. This approach represents therefore a powerful means to bring lung structure and respiratory function into quantitative relation to each other.

Probably the first attempt to correlate quantitatively the function and internal dimensions of the lung was made in 1731 by the Rev. STEPHEN HALES who observed that "the elasticity of the air is reduced ... by 1/3 ... by the respiration of the lungs". Curious to know what happenend inside this organ, HALES set out to study geometry and dimensions of the alveoli of a calf's lung and arrived at the conclusion that they were cuboid boxes "about 1/100 part of an inch in diameter" (254 μ). From this he estimated the "inward surface of the lung [to be] 289 square feet" (27 m²). He concluded that this enormous inner surface of the lung "makes it very probable" that a large part of the "air particles" found in the blood enter it through the lung, rather than from "animal foods", since the blood is, "by an admirable contrivance there spread into a large expanse, commensurate to a very large surface of air, from which it is parted by very thin partitions; so very thin, as thereby probably to admit the blood and air particles ... within reach of each other's attraction."

Since Stephen Hales' time the interpretation of the physical events involved in pulmonary gas exchange between air and blood has been revised on various essential points. Accordingly, it is presently possible to put the quantitative correlation of alveolar structure and respiratory function on a more rational basis. MÜLLER (1945) theorized about the effect of different alveolar dimensions on the interpretation of gas exchange phenomena. For this purpose he availed himself of data found in the literature. However, the great degree of variation of the alveolar measurements obtained by several investigators led him to the conclusion that the dimensions of inner pulmonary structures were yet too ill defined to allow any functionally significant conclusions to be drawn from quantitative physical considerations on diffusion events at the alveolo-capillary level.

Table I.1. *Dimensions of Human Lung in Literature*

Author		Number of alveoli N_{AT}	Alveolar surface S_{AT} m²	Alveolar diameter d_A μ
Huschke/Sömmering	1844	1800 · 10⁶		
Rossignol	1847			200—500
Frey	1859			50—166.7
Aeby	1880	♂404 · 10⁶ ♀322 · 10⁶	45—115	
Zuntz	1882	725 · 10⁶	90	
Kölliker-v. Ebner	1902			160—370
Schulze	1906	150 · 10⁶	30	
Ogawa	1920			100—190
Willson	1922	150 · 10⁶	47—70	75—125
Arthus	1927		200	
Marcus	1928	444 · 10⁶	50	150
Gertz	1928		48	
Claus	1935			300—600
Macklin	1943			166
v. Hayek	1953	300—400 · 10⁶	30—100	200—250
Kulenkampff	1954		32—50	
Elze and Hennig	1956	66 · 10⁶		
Engel	1958			200
Hieronymi	1961	75 · 10⁶	30	300
Radford	1953		5—10	
Clements	1960	5000 · 10⁶	55	60

Table 1 lists the number of alveoli, their diameter and their total surface as they were estimated over the last 120 years by a sequence of authors. It can be seen that the total alveolar surface area was calculated anywhere from 30 to 200 m²; the average alveolar diameter was found to measure from 50 to 600 μ. This wide range of estimates could partially be due to a variation in the degree of inflation of the investigated lungs; however, this variable should not influence the total number of alveoli. Yet, Table 1 shows that this number was estimated from 75 to 725 million, not counting the extreme cases of Huschke (before 1844) with 1800 million, or of Clements with 5000 million, since the latter figure was obtained by indirect means. A large part of these differences can be accounted for by various indeterminates in the methods employed. For example, in most cases the size of the lungs was not actually determined. This is a critical omission. In many instances, the methods of measurement were also questionable except for the most recent studies by Kulenkampff (1954) and Hieronymi (1961) who used the method of mean chord length for an estimation of the alveolar surface area. Their results will be discussed later in further detail.

Is is obvious from this brief review that a careful reinvestigation of alveolar dimensions is needed if any weight is to be ascribed to these quantities in physiologic considerations. This appears to be of particular importance in view of recent comparative studies by Tenney and Remmers (1962) who found, in a large series of mammals, that the relative alveolar surface area per unit lung volume was proportional to the basal oxygen requirement per unit body weight of these animals. This suggests that the efficiency of gas exchange between air and blood may be significantly influenced by alveolar dimensions. Of perhaps even greater importance, however, are the dimensions of the alveolar capillary network which, to our knowledge, have never been quantitated morphologically, although some physiologists have estimated the capillary volume by indirect methods (ROUGHTON 1945; McNEILL et al. 1958; FORSTER 1959; LEWIS et al. 1958; BATES et al. 1960).

In recent years electron microscopy has also allowed an estimation of thickness and composition of the alveolo-capillary tissue barrier separating air and blood (LOW 1952; KARRER 1956; BARGMANN and KNOOP 1956, MEESSEN 1960; and others). In its thinnest portions it was found to measure about 0.4 μ. However, as will be discussed later, it is not satisfactory to consider only the "effective" thinnest portions of the alveolo-capillary tissue barrier (MEESSEN 1960), since prejudice is thus introduced into the reasoning. It is better to estimate the distribution of thicknesses over the respiratory surface.

Between environment and alveoli the air is transported through an elaborate system of airways whose geometry and dimensions may influence this part of respiratory function to a considerable extent. In his classical work on airway resistance ROHRER (1915) measured meticulously the dimensions of the bronchial tree by means of calibrated bougies. This method apparently failed him toward the periphery of the lung, in that he greatly underestimated the number of smaller bronchi. FINDEISEN (1936) studied the deposition of air-borne particles in the human lung. For that purpose he theoretically constructed a dimensional airway model on the basis of scanty information taken from the literature. This model incorporates fundamental errors; it has nevertheless often been used a as source for information on the airways (LANDAHL 1950, Handbook of Respiration 1958, etc.). More recently, Ross (1956) measured the variation in airway dimensions of a dog's lung, and studied the effect of irregularities of the bronchial tree on the distribution of ventilation. However, the applicability of these data to the human lung is limited.

From this brief review it becomes evident that available data on the dimensions of pulmonary structures are incomplete and uncertain. A systematic reinvestigation of these dimensions is therefore needed before a quantitative correlation of lung structure and respiratory function can be attempted on a physico-mathematical basis (GOMEZ 1962). For this reason the present morphometric study was initiated; its goals are summarized in the following pages.

It was first decided to concentrate on the investigation of structural parameters which are of primary significance with respect to gas exchange. We have therefore determined those pulmonary dimensions which characterize the alveolar and capillary respiratory surfaces, the capillary blood volume, and the alveolo-capillary tissue barrier. In addition, we have investigated geometry and dimensions of the airway system which distributes the air into the alveoli. The pulmonary circulation could, so far, only be quantitated in its most peripheral sections since it poses special problems which exceed the scope of this particular study. Auxiliary structures, such as the bronchial circulation, the lymphatic system and the lung nerves were disregarded for the time being. Neither was the fine structure of the various elements studied in detail; this particular subject has been extensively described in the literature (cf. MILLER 1937, HAYEK 1953, KRAHL 1959 and others). Besides a quantitation of fine structural elements would be beyond the goals of this project.

The major part of these studies was carried out on five normal human lungs which were prepared for microscopic investigation. It was recognized that a rigorous standardization of the methods of preparation was essential to avoid such undesirable effects as different degrees of inflation on the dimensions of peripheral lung structures. The method which finally evolved from preliminary studies is outlined in Chapter IV. Additional information was obtained from plastic casts of the bronchial tree, and from bronchograms prepared on the living; this will be discussed in further detail in the appropriate sections.

In order to achieve sufficiently reliable estimates of the various dimensions it was found essential that the procedures of sampling and measuring employed should be both accurate and efficient. Wherever feasible all measurements were obtained on random samples, which allowed the sample size to be reduced to a minimum. The methods of measurement, which varied from task to task, involved, for example, the counting of structures, the analysis of volumetric proportions, the estimation of surfaces, of surface-to-volume ratios, and of some linear dimensions. All these methods operated on sections which are two-dimensional samples of the tissue. The principles on which they were based are extensively discussed in Chapter III. Because of rigorous randomization of sampling and measuring procedures, the 95%-confidence interval attained was always considerably less than 10% of the mean, despite the surprisingly small number of samples used. This is evidently satisfactory for our purposes.

The present monograph will describe our general approach to morphometric analysis. It will discuss the available methods of sampling, measurement and preparation as they apply to the lung in particular, and it will then present the findings obtained in normal human lungs. Wherever feasible the various dimensions will be related to each other on the basis of geometrically derived formulas. Finally, the formulation of geometric and dimensional models of the human lung, which should facilitate the application of these findings to functional considerations, will be attempted.

Chapter II

Organization of the Human Lung

The lung[1] contains three basic components: air, blood and tissue. Air and blood are transients and are moving in and out of the organ at a quite rapid rate. The tissue, however, is comparatively stationary. It confines air and blood to separate compartments, forms channels which are suited to lead the air in and out of the lung, or to allow the blood to flow through it. The tissue framework of the lung also insures a constant adequacy of the intricate relations between air and blood. Later on, the architecture of the peripheral parts of the lung will be

---

[1] The term "lung" (singular) refers to the entire "organ lung", that is, to the joint right and left "lungs", since we shall not be interested, at first, in topographic differences. Whenever an uncertainty arises, we shall speak of "entire lung" and of right and left "half-lungs".

compared to a fine froth (Fig. 39) in which gases are exchanged between the air bubbles and the blood film surrounding them. An important difference between froth and lung is brought about by the presence of a well organized tissue framework in the latter, which allows the air to be freely exchanged between the bubbles and the environment, and which lets the blood flow continuously through the films, without the necessity of breaking the "froth".

Part of the tissue forms the wall of airways and blood vessels; these sheaths are responsible for maintaining continuity and proper dimensions of their respective compartments. They can also act in various ways upon the content through specially differentiated structures, such as smooth muscle fibers, or glands, to which much attention has been given in morphologic research on the lung (MILLER, 1937; VON HAYEK, 1953; TÖNDURY and WEIBEL, 1958; and many others). Another part of the lung tissue is interposed between these channels, and molds them into a coherent organ.

The close relationship between the lung structures can be appreciated by studying early stages in the development of the lung. Fig. 1a shows a parasagittal section of the chest of a human embryo of about three weeks. The thoracic cavity contains a mass of loose mesenchyma, which is bounded by mesothelium and, already at this early age, reveals the basic shape of the lung. Airway channels, which are derived from the gut, are seen to sprout into this mesenchymal anlage of the lung. They are accompanied by branches of the pulmonary artery. The still abundant mesenchyma contains precursors of the capillary network and branches of the pulmonary vein. With further development (Fig. 1b) the relative amount of mesenchyma is gradually reduced by proliferating airways and blood vessels to a final minimal interstitium, which contains a highly organized framework of connective tissue elements (ORSOS, 1907; Low, 1952; KARRER, 1956; LOOSLI, 1959; and others).

In the mature lung, airways and blood vessels have a characteristic relationship to each other (Fig. 2): the pulmonary arteries are seen to accompany the airways quite closely, while pulmonary veins are situated between two airway trees. This arrangement is maintained throughout the lung, from the hilum out to the periphery. Based upon it has come the definition of a sequence of broncho-vascular units such as pulmonary segments, lobules, and so forth (WINKLER, 1952; TÖNDURY, 1956; and others). In part this was but a further refinement of the classical subdivision of the two lungs into lobes, segments, lobuli and acini (Fig. 3) (MILLER, 1937; VON MÖLLENDORF, 1941; JACKSON and HUBER, 1943; VON HAYEK, 1953; ENGEL, 1958; and others).

Although this classic concept allows a clear anatomic description of the lung, at first, it does not conveniently lend itself to our approach, which will attempt to analyze the internal architecture of the lung and define the dimensions of its elementary components. We were guided in our study by the specific requirements of morphologic knowledge in physiologic reasoning on respiration. From this point of view we shall consider the lung, in first approximation, as constructed of four separate compartments (Fig. 4a): A: Airways, B: Pulmonary arteries, C: Pulmonary veins, and D: Lung tissue. The walls of airways and blood vessels will be part of the tissue compartment D and will not be discussed here, except for the tissue barrier at the alveolo-capillary level.

As compared with the arrangement of airways and blood vessels in the actual lung (Fig. 3), it may appear that the relationship of these structures to each other was inverted in the model of Fig. 4a. While the origins of the three compartments (airways, arteries and veins) seem to lie in the center of the organ, they have now been shifted to its periphery. This is true in a topographic sense. Functionally, however, the center of the organ is located at its topographic periphery where the airways terminate in the alveoli and come into close relationship with the capillary

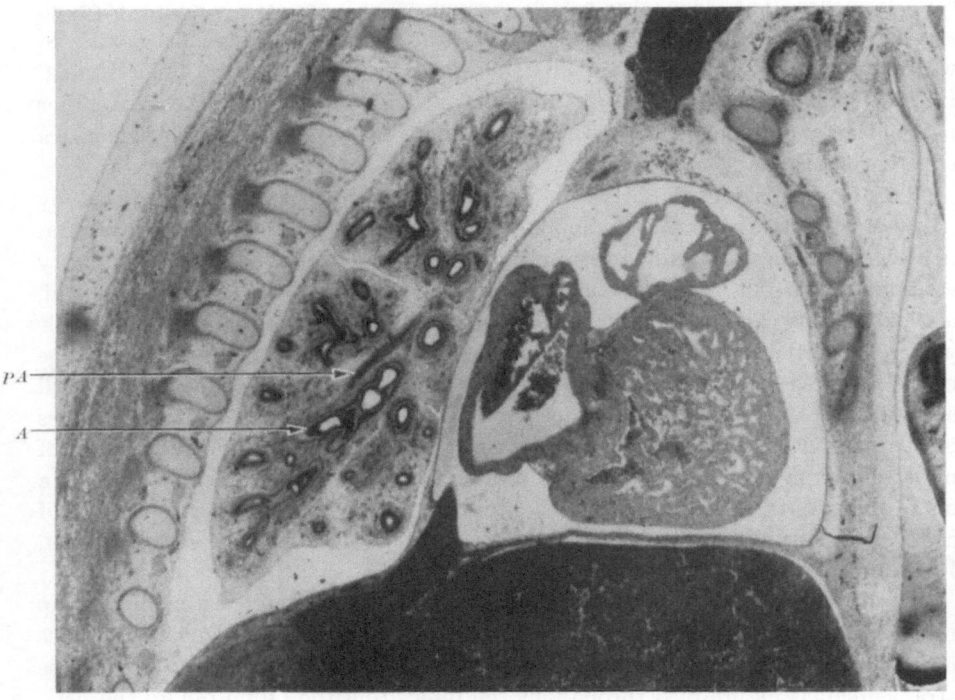

Fig. 1a. Parasagittal section of chest of human embryo of 22 mm cephalocaudal length. Anlage of lung shows abundant mesenchyma in which the primitive airways (*A*) and the blood vascular systems are embedded. *PA* = branch of pulmonary artery. × 23

network which joins arteries and veins. In this sense, the model shown in Fig. 4a is well able to represent the functional "topography" of the lung, in particular that of one lung "unit". The entire organ would then be composed of many millions of such triangular units arranged approximately parallel to each other, with the three vertices merging in one point (Fig. 4b).

In reality, these functional lung units do not arise independently from the origin of the three functional compartments. Their large number is rather gradually attained by a sequence of branching illustrated schematically in Fig. 5. Each compartment can, therefore, be subdivided into a large number of geometrically simple elements. These are arranged in a tree-like series-parallel system (Figs. 2, 3 and 5) which distributes air and blood gradually into the functional units. A typical element of the airway system would be represented by a roughly cylindric bronchial segment, which stretches from its proximal to its distal branching site (Fig. 5). A vascular element would have similar characteristics.

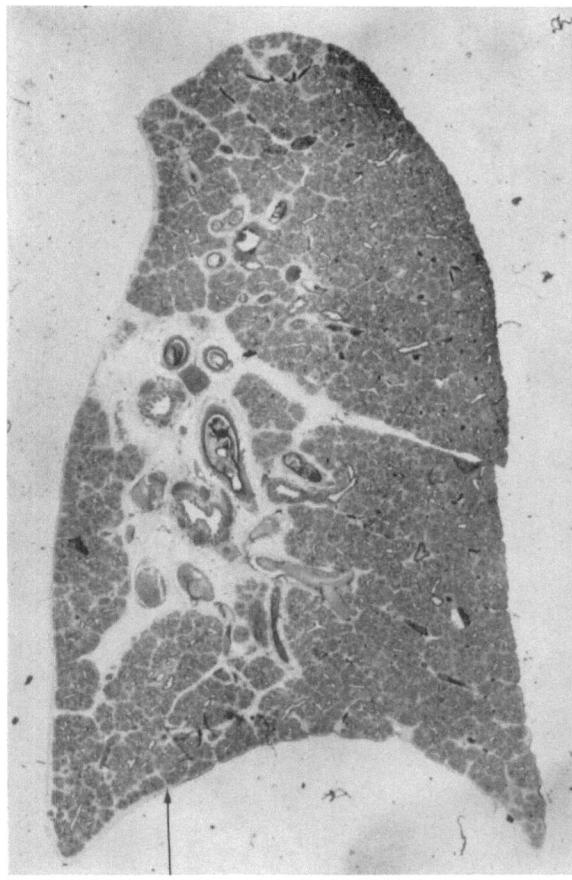

Fig. 1b. Frontal section of dorsal portion of left lung of human embryo of 57 mm cephalocaudal length. Airways and blood vessels have multiplied extensively by ramification. In the periphery, mesenchyma occupies smaller fraction of lung volume in form of fine septa (S). × 6.5

Fig. 2. Peripheral portion of plastic cast of bronchi (B), pulmonary artery (PA), and pulmonary vein (PV) of human lung showing the characteristic topographic relationship of these elements (from TÖNDURY and WEIBEL, 1958)

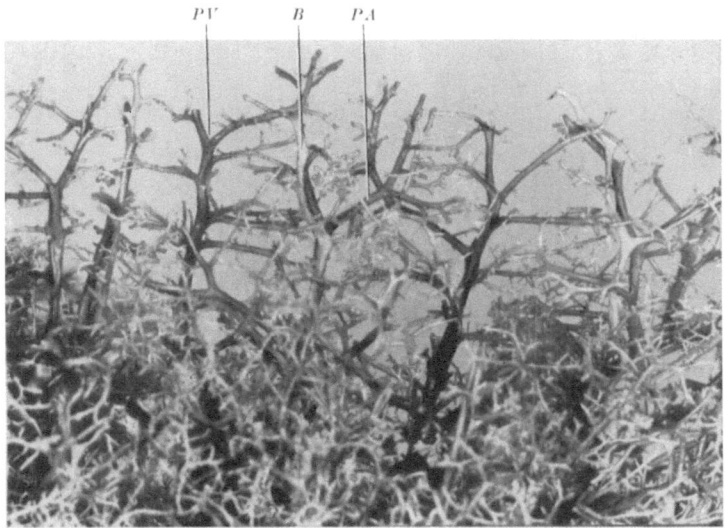

Fig. 2

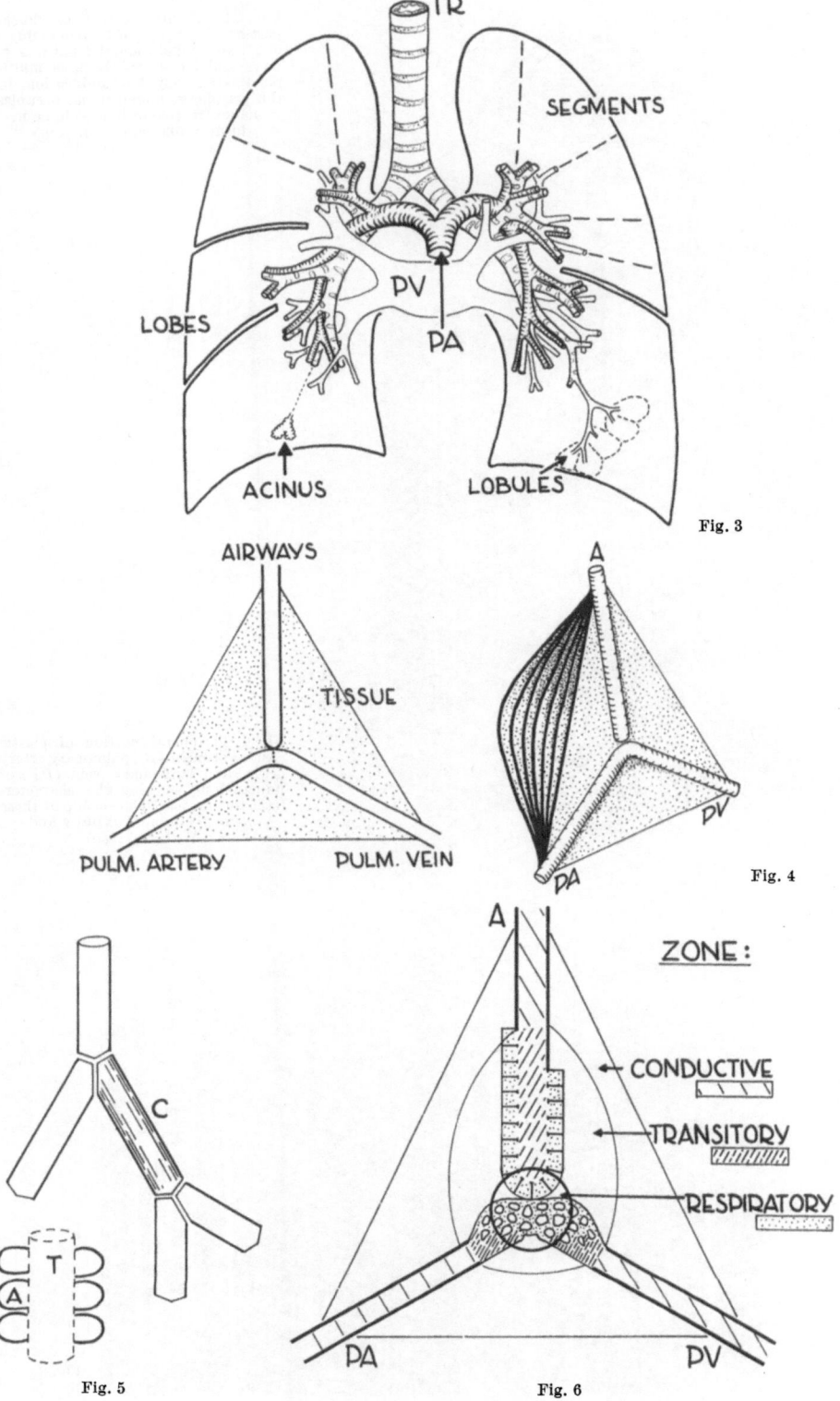

Fig. 3

Fig. 4

Fig. 5

Fig. 6

On the other hand, the terminal airway elements are represented by alveoli which are lateral outpouchings of alveolar ducts.

Structural and functional characteristics will allow us to assign these elements to three concentric zones (Fig. 6) for which we may anticipate different types of functional behavior:

The *respiratory zone* contains the alveoli and the alveolar capillaries, which are in such intimate contact that the exchange of gases between air and blood can effectively take place.

In the elements of the *conductive zone*, air and blood are enclosed by a well developed wall, which may exert some regulatory influence on its content. This zone includes bronchi, bronchioles, pulmonary arteries and veins.

In the *transitory zone* elements of the respiratory zone are intimately connected with elements of conducting character. Alveoli are apposed to respiratory bronchioles, alveolar ducts and sacs; pre- and postcapillaries lead directly into the capillary network.

In the following analysis we shall investigate separately the elements of these three zones. Finally we shall attempt to integrate them in the form of an over-all model of the lung.

Chapter III

Principles and Methods of Morphometry

For obvious reasons organs of the size of the human lung cannot be completely investigated. Every study of the properties of its internal structures will have to be carried out on samples of the tissue, particularly on sections which can be analyzed microscopically. In descriptive morphology it is customary to sample typical regions which are then carefully investigated. This procedure, however, gives no indication of the frequency and distribution of the structures analyzed. It may, thus, lead to an unjustified overestimation of the importance of some peculiarities of the tissue and to a misinterpretation of their significance. For example, the occurrence of internal longitudinal muscle bundles in systemic arteries of the lung was interpreted as a regulating mechanism ("Sperrarterie") because it was found in connection with anastomoses between bronchial and pulmonary arteries (v. HAYEK, 1940; VERLOOP, 1948; and others). A systematic study of the distribution of such arteries with longitudinal muscle bundles, however, revealed that they were only rarely associated with broncho-pulmonary anastomoses but occurred commonly all along the course of the bronchial arteries (TÖNDURY and WEIBEL, 1956 and 1958; WEIBEL, 1958 and 1959). This led to

---

Fig. 3. General topography and organization of lung structures. Lobes, segments, lobules and acini are lung units of classic anatomy

Fig. 4a and b. Functional lung units have center in alveolo-capillary region where airways and blood vessels converge. The lung is composed of many such units which are arranged parallel with common origins of airways and vessels

Fig. 5. Airway elements of conductive (C), transitory (T), and respiratory zones (A = alveoli)

Fig. 6. Concentric arrangement of conductive, transitory and respiratory zones of airways (A) and blood vessels (PA and PV)

a re-interpretation of the longitudinal muscle bundles as a reaction of the vessel to intermittent stretching by the respiratory movements of bronchi; this hypothesis could be substantiated experimentally (WEIBEL, 1958). Such misinterpretations due to inadequate sampling procedures may be encountered even more frequently in pathologic studies, in which it is quite customary to study typical lesions.

It is immediately obvious that this arbitrary approach is not adequate for the morphometric studies pursued here. In establishing working rules for morphometry we shall need to direct particular attention to the question of adequate sampling, since it will always be necessary to obtain an estimate of the raletive frequency and of the distribution of the structures under investigation.

For this purpose we have two fundamental sampling methods at our disposition: random and systematic sampling. Random sampling is the most efficient and also the most reliable procedure and will be used wherever the structures are randomly or homogeneously distributed (see below). Systematic sampling is usually more laborious, but may be necessary in analyzing non-randomly distributed structures, or as a subsampling procedure within random samples.

The methods of measurement employed in morphometry depend on the characteristics of the structures under investigation, on the information required, and on the sampling procedures used. In general, they should efficiently furnish a comparatively large amount of information which is amenable to statistical analysis. Most methods of systematic measurement are very laborious, and often require time-consuming, three-dimensional reconstructions. They, thus, allow only a small sample of tissue to be investigated in available time. We will, therefore, attempt to use methods involving random measurements, whenever this is feasible. Some of these methods were already available at the start of this study; some were developed in its course. Many more need to be worked out in due time.

In subsequent paragraphs we will discuss the special aspects of two-dimensional random sampling and of systematic sampling of the human lung, and propose practical procedures. We will, then, present and critically evaluate morphometric methods of measurement which are related to this work.

A. Fundamental Questions of Sampling for Morphometric Studies

1. Randomness and Non-randomness of Distribution of Structures

Morphometry makes extensive use of statistical procedures. By its nature statistics, in most instances, asks for random conditions in the substrate analyzed. It may, therefore, appear impossible to apply statistical procedures of analysis to biologic structures which possess an extremely high degree of organization. In fact, the results of these studies will be used to demonstrate a certain part of this organization rather conclusively. In view of this apparent incompatibility of concepts, it is necessary to define how "randomness" of distribution of structures is understood in this study.

Pulmonary alveoli, for example, are organized rather rigidly into units of higher order, alveolar ducts or sacs, by being arranged around a central duct as will shortly be described. The central ducts, on the other hand, are part of the well organized system of airways. There is, however, no lattice which determines the spatial relation between alveoli belonging to different alveolar ducts or sacs. Within 1 cm³ of parenchymatous lung tissue, for example, alveoli can be considered randomly packed, since this unit volume will contain many units of higher order. Another way of defining this random distribution is to demand that sections of the tissue in all directions of space produce the same picture. This condition is evidently satisfied with respect to alveoli except perhaps in such rare locations as in the immediately subpleural zone.

As another example, a renal glomerulus is in well organized relation to its nephron; glomeruli of different nephra, however, are randomly distributed within the cortex of the kidney. But with respect to the entire kidney, glomeruli are not randomly distributed since the cortex is present only in well confined regions (cf. ELIAS et al., 1961). In reference to the lung, alveoli are randomly distributed only with respect to the respiratory lung tissue or parenchyma, but not with respect to the entire lung, since the conductive zones of airways and blood vessels are clear examples of non-randomly distributed structures. In the conductive zone, we find a spatial sequence of structures which are not equivalent to each other. The centrifugal tree-like branching of airways, for example, imposes a spatial lattice on the arrangement of conductive airway elements which cannot be ignored. Another simple example of non-random distribution is found in the cells of a stratified squamous epithelium where, obviously, sections in different directions of space will yield very different pictures.

We arrive at the following conclusions:

a) Random distribution of structures in space can be assumed if the units under investigation do not exhibit any stratified array in the unit tissue volume, even though they may be well organized into units of higher order. In other words, sections in all directions of space must yield identical pictures.

b) Randomness of distribution of structures refers only to a specified part of the tissue; alveoli, for example, are only randomly distributed with respect to lung parenchyma.

c) It is necessary to specify the degree of randomness of distribution for every structure under investigation.

d) If randomness of distribution cannot be assumed, ordinary statistical procedures cannot be employed. It will be necessary to define the underlying lattice and to devise appropriate statistics.

2. Sampling of Randomly Distributed Structures

a) Justification of Two-dimensional Sampling of Tissues

The investigation of the internal structure of an organ must be carried out on tissue sections. For qualitative studies it has been intuitively assumed that such sections are representative samples of the tissue. However, if tissues should be investigated quantitatively by means of tissue sections, the fundamental question arises whether sections can represent the tissue structures quantitatively as well.

A tissue section is a very thin slice of tissue whose thickness, for technical reasons, is reduced as far as possible in order to obtain good lateral resolution. For most purposes, it has been found best to cut sections of 7 μ for light microscopy, and of 600—900 Å for electron microscopy. Since most quantitative procedures described hereafter use comparatively low magnifications, the section thickness can be sonsidered very small, so that it can be disregarded in most instances. A section can, therefore, be considered to be a two-dimensional sample of the tissue. This circumstance is most convenient for morphometric purposes, since it allows us to confine all quantitative procedures to a determination of lateral dimensions, which is usually easy. The determination of dimensions along the optic axis, on the other hand, is rather cumbersome and often unreliable, so that it is very fortunate that depth need not be considered.

In view of this circumstance the introductory fundamental question can be rephrased more precisely: *Can a two-dimensional sample of a three-dimensional system of structures (tissue) be quantitatively representative of the composition of the tissue?*

The answer to this question is given by the principle developed by the geologist DELESSE in 1842 which states: If a section is placed through a tissue volume containing a given component, the fraction of the area covered by transsections of the component will be equal to the fraction of the volume occupied by the component. Since this principle is of such fundamental importance for the entire morphometric approach pursued in this study, its formal demonstration will be attempted.

Suppose that a cube (Fig. 7) with volume

$$V = L^3 \tag{3.1}$$

contains granules of any shape and size which together have a volume

$$v = \varrho V . \tag{3.2}$$

Consider, now, a thin slice of this cube of thickness dx parallel to the (z, y) plane having a volume

$$dV = L^2 \cdot dx . \tag{3.3}$$

In this slice a volume

$$dv = \eta(x) \cdot dV = \eta(x) \cdot L^2 \cdot dx \tag{3.4}$$

will contain segments of the granules. L and dx are constant wherever we place the slice but $\eta(x)$ will vary with x, as is indicated in Fig. 8. If $dx \to 0$ the total volume of the granules v is

$$v = \int_0^L dv = L^2 \cdot \int_0^L \eta(x)\, dx = \varrho \cdot L^3 . \tag{3.5}$$

But

$$\frac{1}{L} \int_0^L \eta(x)\, dx = \bar{\eta} \tag{3.6}$$

is the average value of the coefficient $\eta(x)$ between 0 and L so that it follows from (3.5) and (3.6) that

$$\bar{\eta} \cdot L^3 = \varrho \cdot L^3$$

or

$$\bar{\eta} = \varrho . \tag{3.7}$$

Relation (3.7) means that the volumetric frequency ϱ of a given tissue component is reflected on sections of the tissue in occupying a corresponding fraction η of the section area. In more practical terms, an average fractional coefficient $\bar{\eta}$ determined on sections through a volume represents an estimator of the volumetric frequency ϱ of the component under investigation. From this the conclusive answer to the introductory question follows: *A tissue section is a quantitatively*

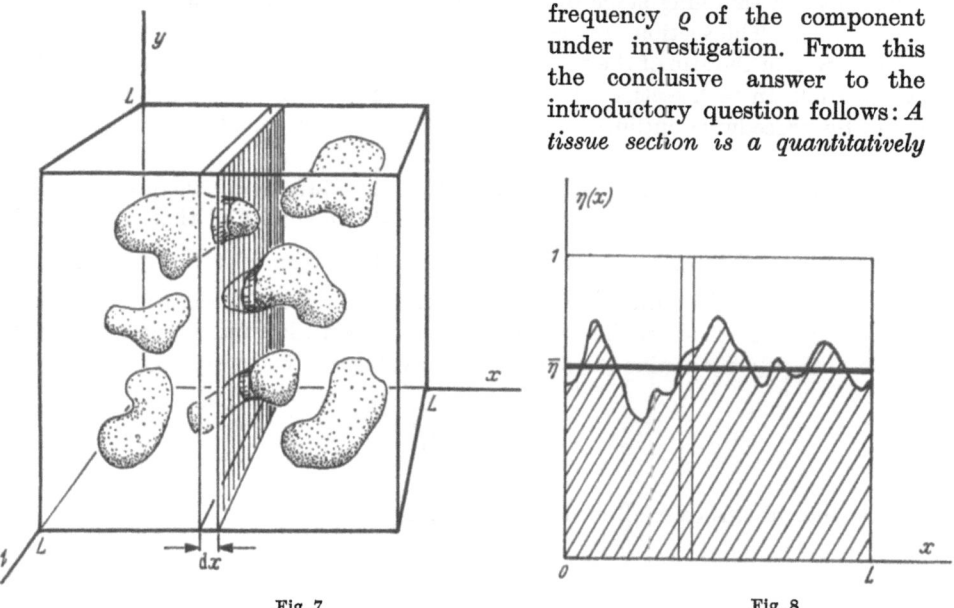

Fig. 7

Fig. 8

Fig. 7. Model for demonstration that random section is faithful representation of tissue composition. Compare text

Fig. 8. Derivation of Delesse principle. Variation of planimetric fraction $\eta(x)$ around average $\bar{\eta}$. Compare text

[From WEIBEL 1963]

representative two-dimensional sample of a three-dimensional system of randomly distributed tissue structures. The statistical rules testing the significance of this sample estimation will be discussed later. They will depend on various characteristics of the component under investigation, for example, on their size, geometry and degree of dispersion.

b) Two-stage Sampling for Microscopic Work

Many structures which are of particular interest are of such small dimensions that they must be studied by either light- or electron microscopy. These methods allow the investigation of only very small areas in one step. Preparative procedures, on the other hand, require that larger pieces of tissue be processed, as will be discussed. Our measurements are, therefore, obtained on samples acquired in two stages.

In the first sampling stage we operate on the fixed lung. (In other organs this may also be done on the fresh specimen.) We consider the lung subdivided into N blocks of 3 to 5 cm³ in volume, and draw a simple random sample as outlined on p. 45. In an adult lung the total number of blocks is of the order of $N \approx 1000$; in general, a sample of $n = 5$ to $n = 10$ units will be sufficient. It, therefore, follows from the combinatorial formula that the number of possible samples is very large:

$$\binom{N}{n} = \frac{N}{n!\,(N-n)!} \approx 10^{12}\,. \tag{3.8}$$

Simple random sampling requires that each of the $\binom{N}{n}$ possible samples have the same chance of being chosen. On p. 45 we shall propose two practical procedures by which simple random samples can be procured. The blocks thus obtained are processed for histologic preparation and sectioned. In the preceding paragraph it was shown that two-dimensional sections are representative samples of the tissue block, if the structures under investigation are randomly distributed.

The second sampling stage operates on the unit sample sections obtained in the first stage. The section area is here considered to be covered by a total of M subunit areas of the size of the microscopic field, which is needed for the particular investigation. In each section a simple random sample of m subunit fields is again chosen by procedures outlined on p. 47, and the measurements are obtained on these subsamples.

This two-stage sampling procedure defines the statistical analysis to which the measurements must be subjected. In general, we shall find that $n \ll N$ and $m \ll M$, so that the ratios $\frac{n}{N}$ and $\frac{m}{M}$ are negligible. A correction for finite population can, therefore, be omitted. By the symbol y_{ij} we denote an observation made on the jth subunit (field) of the ith unit sample (section). For most purposes it will be satisfactory to determine the over-all mean value of y with respect to subunits:

$$\bar{\bar{y}} = \frac{\sum\limits_{i=1}^{n} \sum\limits_{j=1}^{m} y_{ij}}{n \cdot m} \tag{3.9}$$

The variance of this over-all mean is estimated by

$$s^2(\bar{\bar{y}}) = \frac{\sum\limits_{i=1}^{n} (\bar{y}_i - \bar{\bar{y}})^2}{n(n-1)} \tag{3.10}$$

where

$$\bar{y}_i = \frac{\sum\limits_{j=1}^{m} y_{ij}}{m} \tag{3.11}$$

is the mean of all observations in the ith unit. For further information on the statistics of two-stage or multiple-stage sampling we refer, for example, to COCHRAN, 1953.

c) Evaluation of Sample Size

In a study of this complexity no general rule for the evaluation of the sample size can be given, since it depends on various circumstances particular to the investigative procedures used at each step, and on the precision required. In general, the sample size had to be determined rather arbitrarily, since it was obviously not easy to define a cost function which would be necessary for an unambiguous evaluation of the required sample size for specified confidence limits. But the purpose of this study was to obtain, in a workable period, a large amount of information covering a wide spectrum, rather than a high degree of accuracy on a limited number of points. On the whole, we were, therefore, satisfied to obtain 95% confidence intervals of $\pm 10\%$ of the mean.

The half-width d of the confidence intervals was, therefore, determined secondarily from the data according to the usual formula

$$d = \frac{t \cdot s}{\sqrt{n}} \qquad (3.12)$$

where s is the standard deviation of the mean and n the number of samples. The coefficient t is defined as the deviation of the sample mean $\bar{y}$ from the population mean μ, given by the equation

$$t = \frac{\bar{y} - \mu}{s/\sqrt{n}} \, . \qquad (3.13)$$

Its value is given in Student's t-table for the specified confidence limits and for the given degrees of freedom $(n - 1)$.

As already pointed out, peripheral lung structures were measured microscopically on two-stage samples. Alveoli, alveolar ducts and capillaries are quite homogeneously distributed throughout the lung. It was, therefore, found that a primary sample of five sections was sufficient to satisfy the specified confidence requirements provided that an appropriate number of subsamples was chosen.

If greater accuracy is required, or if the structures under investigation are inhomogeneously distributed, it will be necessary to increase the number n of primary samples. Little would be gained by increasing the number m of subsamples, since a larger n can more effectively reduce the variance of the over-all sample (cf. equation 3.29). This may be particularly important in applying these morphometric methods to pathologic specimens. If the inhomogeneity is marked, or if it is a characteristic property of the disease, it will be advisable to draw stratified random samples rather than simple random samples (see e. g. COCHRAN, 1953, p. 65ff.).

3. Sampling of Non-randomly Distributed Structures

Some of the lung structures which will be examined subsequently do not satisfy the conditions for randomness of distribution as defined (on p. 10). For example, conductive structures will be found in greater density in the hilar regions than in the periphery of the lung (Fig. 33). In addition, the dimensions of conductive structures change significantly from the hilum toward the periphery. This last property is an important characteristic of conductive airways and blood vessels and will be one of the main features to be analyzed. It is, therefore, obvious that random sampling procedures, as described in the preceding paragraphs, cannot be used at this stage of the analysis. Rather, we shall need to design suitable schemes of systematic sampling which will permit us to obtain the required information.

On the one hand, parenchyma[1] and non-parenchymatous structures are not randomly distributed. But their relative fractional volumes can nevertheless be determined by point-sampling on a hexagonal point-lattice superimposed on equidistant cut surfaces of the fixed lungs (Fig. 33). The distance of the cut surfaces corresponds approximately to the distance of the points, so that the entire procedure amounts essentially to suspending a three-dimensional tetrahedral

---

[1] For a definition of the term "parenchyma" as used in this context, see footnote on p. 52.

point-lattice in the lung. Since all cut surfaces of the parallel lung slices were covered completely with the point-lattice, the entire lung is sampled systematically at equidistant points in space. The number of points necessary for this analysis depends on the magnitude of the fraction assessed, and on the acceptable error, as will be discussed on p. 21.

The dimensions of the elements of the airway tree had to be analyzed in such a fashion that the change in dimensions, progressing with distance from the origin, became apparent. Since airways were found to branch by dichotomy, it was possible to design a procedure of systematic sampling by stratification with respect to generations of branching. Sampling each generation as completely as possible was attempted, since it was not possible to employ either random or systematic sampling methods within these strata. Where completeness was not achieved, the sample was significantly biased. However, it was possible to correct this bias by appropriate statistical treatment of the data. Because of its special character this analysis will be discussed in Chapter X, p. 115.

The alveolar capillary network shows an intermediate distribution pattern. Its elements are non-randomly arranged in a two-dimensional network which is spread within single facets of the interalveolar septa (Fig. 49). The facets, however, can be assumed to be randomly distributed in space according to the above criteria. In order to quantitate the capillary network, a simple random sample of interalveolar facets was drawn; in each of these facets all capillary elements were then systematically measured.

B. Principles and Methods of Microscopic Quantitation on the Basis of Two-dimensional Samples

The principle of DELESSE, which was presented on p. 12, demonstrates that sections are quantitatively representative two-dimensional samples of tissues containing "randomly" distributed structures. It justifies the use of random sections in estimating quantitatively various parameters relative to these structures. This will involve the determination of volumetric proportions of tissue components, the counting of the number of certain structures, and the estimation of some of their dimensions.

1. Principles for Analyzing Volumetric Proportions of Tissue Components

a) The Linear Integration Method

Extending the principle of DELESSE (p. 12), the geologist ROSIWAL demonstrated in 1892 that the fraction of a line passing through a randomly distributed tissue component is approximately equal to the fraction of the volume occupied by this component. A straight line passing through the tissue can be considered as lying on a plane section of the tissue. If a line is, therefore, laid on a section of the tissue, the fraction φ_G of this line lying on transsections will be an estimator of the fraction η_G of the area covered by transsections. The demonstration of this relation follows a path similar to that pursued in deriving Delesse's principle.

Let a square area $S = L^2$ be covered by spots (transsections of the structure G under investigation) as illustrated in Fig. 9a over an area

$$s_G = \eta_G \cdot S \tag{3.14}$$

and consider a strip of width dx and area

$$dS = L\,dx.\tag{3.15}$$

A fraction

$$ds_G = \varphi_G \cdot dS = \varphi_G \cdot L\,dx\tag{3.16}$$

of this strip will thus be covered by spots. We shall now find that $\varphi_G(x)$ again varies with x (Fig. 9b). If $dx \to 0$

$$s_G = L\int_0^L dx \cdot \varphi_G(x) = \eta_G \cdot L^2.\tag{3.17}$$

But since

$$\frac{1}{L}\int_0^L dx \cdot \varphi_G(x) = \bar{\varphi}_G\tag{3.18}$$

is the average value of $\varphi_G(x)$ between 0 and L, it follows that

$$\bar{\varphi}_G = \eta_G.\tag{3.19}$$

In deriving Delesse's principle, it was shwon that the average value of

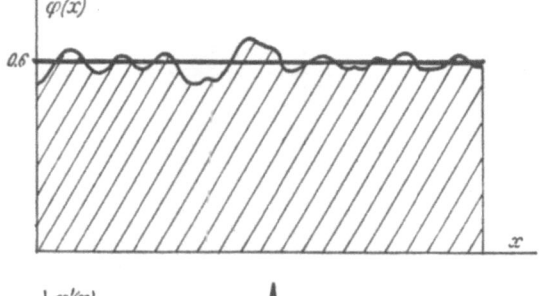

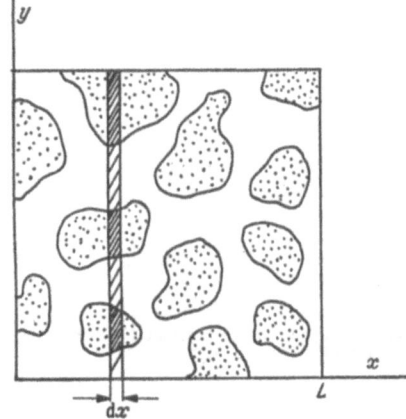

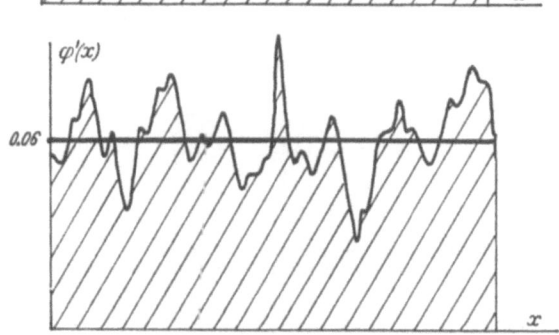

η_G is an estimator of the volumetric fraction ϱ_G occupied by the structures under investigation. We, therefore, find that

$$\varrho_G = \bar{\eta}_G = \overline{\bar{\varphi}_G}.\tag{3.20}$$

Fig. 9a
Fig. 9b
Fig. 9a. Demonstration that volumetric proportions can be determined along random lines (ROSIWAL, 1898). Compare text
Fig. 9b. Examples show that larger scatter of measurements is to be expected if volumetric fraction of component is small (6%), than if it is large (60%) [From WEIBEL, 1963]

ROSIWAL also demonstrated that the line along which φ_G is determined need not be straight but may have any shape, so long as its course is not biased by the underlying array of transsections.

Practical Application of the Linear Integration Method

It is obvious that the principle of ROSIWAL has many practical advantages over the original principle of DELESSE. The measurement of fractions of the length of a line is much less laborious than the planimetric determination of surface fractions. The development of mechanical computing devices (integrating stages) by SHAND (1916), WENTWORTH (1923) and SCHEUMANN (1931) has allowed geologists to analyze successfully the mineral composition of rocks by this principle. In 1948 the pathologist GERSTEL applied the integrating stage to the analysis of histologic sections. Integrating stages, however, lack accuracy, mainly when higher magnifications are to be used. The recording spindles drive the stage and displace the specimen itself along a hairline cross in the eyepiece; this arrangement is very sensitive to slight movements.

SCHUCHARDT (1954) has, therefore, devised an "integrating eyepiece" which is manufactured by E. Leitz, Wetzlar. In this instrument (Fig. 10) a point (hairline

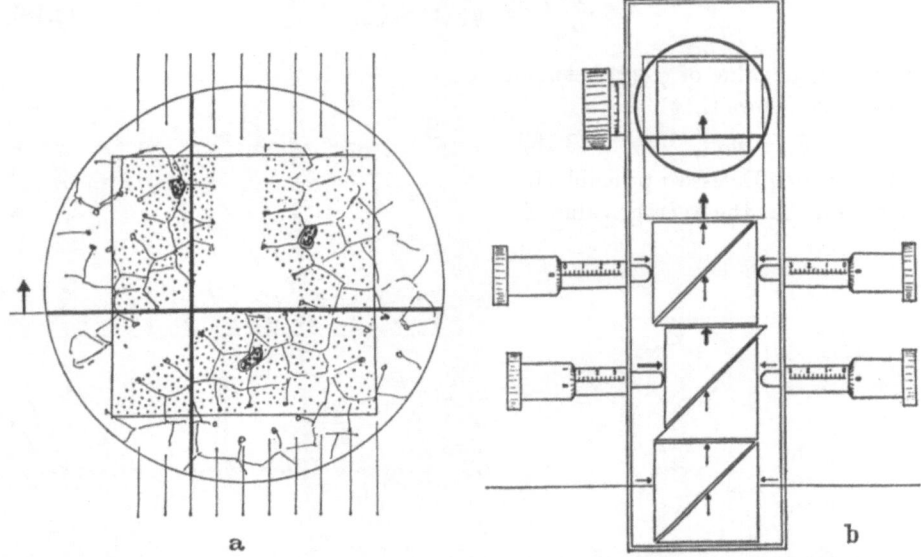

Fig. 10. Integrating eyepiece of Leitz (SCHUCHARDT, 1954). Recording spindles move horizontal hairline along measuring line. Position of hairline cross in diagram is on alveolar duct air. Compare text
[From WEIBEL, 1963]

cross) is moved vertically along a measuring line of fixed length. It is driven over a sequence of wedges by six independent recording spindles, to each of which one given tissue component is assigned. The corresponding spindle is operated when the point passes through a certain component; it will cumulate all individual lengths pertaining to this component without requiring further recording until a series of five measuring lines has been surveyed. The measuring line is displaced by a horizontal spindle. At the end of operation the cumulated lengths L_G are read off each spindle. The linear fractions φ_G are obtained by the relation

$$\varphi_G = \frac{L_G}{\sum_G L_G} \tag{3.21}$$

where $\sum_G L_G$ is the total length of the measuring line.

Application of the Integrating Eyepiece to the Volumetric Analysis of the Lung Parenchyma

As has been discussed, the components of the respiratory and transitory zone of the human lung — alveoli, alveolar ducts, tissue and small blood vessels —

satisfy the conditions defining "random" distribution. The linear integration method could, therefore, be used to determine the relative volumes of these components. Combining the Leitz integrating eyepiece with a plane Fluotar 6x objective, the measuring line measured 0.119 cm. This line was placed at 10 equidistant positions across the square outlined by the eyepiece. Readings of the L_G were taken after five lines had been scanned, yielding two independent readings per field.

From the original set of slides prepared for each case, five slides were drawn at random as described below. By random displacement of the mechanical stage, two fields were subsampled and analyzed on each slide. This procedure yielded four subsamples in each of the five samples analyzed. Fields containing bronchi, bronchioles or vessels larger than 60 μ were discarded before measurement and replaced by new fields, since these structures were not considered part of the respiratory portion of the lung (cf. p. 52).

b) The Point Counting Method

The principles of DELESSE (1847) and ROSIWAL (1892) can be further extended to a procedure of point counting. If a set of points is thrown at random on a section of a tissue containing a component G in a fraction ϱ_G of its volume, the fraction of points lying on transsections of G is an estimator of ϱ_G.

This fact has long been recognized; it is readily demonstrated if we examine the probability of a point lying on transsections of G:

$$\text{Prob}\,(pt \text{ lying on } G) = P_S^G\,. \tag{3.22}$$

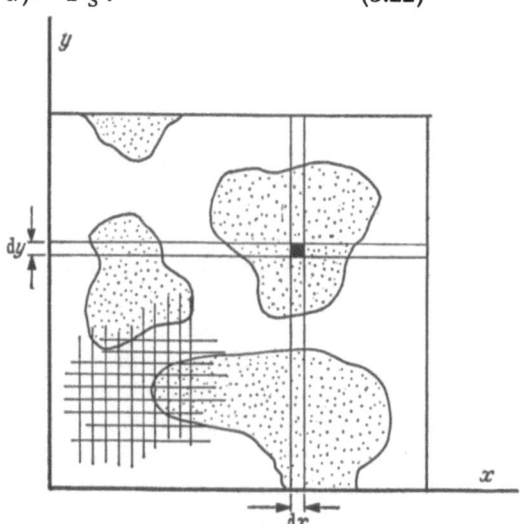

For this purpose let a point be a very small square of area $dy\,dx$. If we now cover the entire field of observation of area S with such points (Fig. 11) the total number of possible squares is

$$n_s = \frac{S}{dx\,dy}\,. \tag{3.23}$$

Of these squares a total number of

$$n_G = \frac{S_G}{dx\,dy} = \frac{\eta_G S}{dx\,dy} \tag{3.24}$$

will lie on granules where $S_G = \eta_G S$ is the area covered by transsections of G. It follows immediately that the probability of a point lying on a granule is

$$P_S^G = \frac{n_G}{n_s} = \frac{S_G}{S} = \eta_G\,. \tag{3.25}$$

Fig. 11. Derivation of principle of point counting for volumetric analysis of tissue composition. Compare text [From WEIBEL, 1963]

By an analogous procedure it is evident that the probability of a point lying in a granule contained in a three-dimensional system of volume V is

$$P_V^G = \frac{n_G}{n_V} = \varrho_G \tag{3.26}$$

since

$$n_V = \frac{V}{dv}$$ (3.27)

and

$$n_G = \frac{\varrho_G \cdot V}{dv}$$

where dv is the shrinking volume of the point. Since we had shown in (3.7) that

$$\bar{\eta}_G = \varrho_G$$

it follows that

$$\overline{P_S^G} = P_V^G = \varrho_G$$ (3.28)

which indicates that the average probability of a point falling onto a transsection of the component G on a section is given by the volumetric fraction ϱ_G of the tissue occupied by G. If we now deposit a number of points n_s on a section and count the number n_G of points on G, we can estimate ϱ_G from equation (3.25).

Practical Application of the Point-Counting Principle

The point-counting principle was first applied to a volumetric analysis of three-dimensional systems by the geologist GLAGOLEFF in 1933. Its application to the volumetric study of biologic tissues at the microscopic level was proposed by

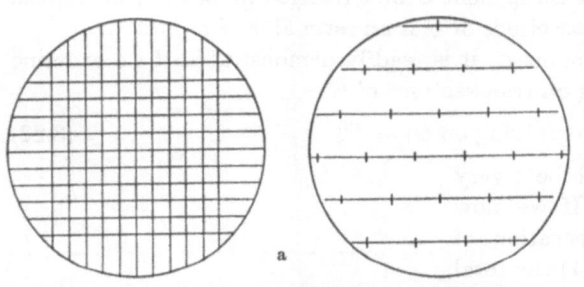

CHALKLEY in 1943 who used a system of five points in an eyepiece. While, in practice, this method is rather cumbersome, the two methods proposed by HAUG (1955) and HENNIG (1956) make efficient point-counting feasible. They differ mainly in number and distribution pattern of the points which are incorporated in the eyepiece of the

Fig. 12. Lattices of points incorporated into eyepiece for point counting volumetry. a 121 points in square lattice according to HAUG (1955), b 25 points hexagonal lattice according to HENNIG (1956) (Integrating Eyepiece I of Zeiss)

microscope. HAUG distributed 121 points in a quadratic lattice while HENNIG used 25 points in a hexagonal lattice (Fig. 12). The choice between the two methods is principally a matter of personal preference. It appears to us, however, that Hennig's proposal has most advantages: the hexagonal lattice provides a more homogeneous distribution of points than a quadratic array; the use of only 25 points on one field is more convenient for counting. It should be noted that the resolution or accuracy of the method is not determined by the number of points counted on a single field but rather by their total number and their distribution. From a theoretic point of view a single point on each field would be ideal since it would allow distribution of the points over a larger area of the section; in practice, however, this procedure would prove tedious. On the other hand, the counting of a large number of points on one field repeated only a few times may well resolve the composition of the particular fields to a high degree of accuracy but the result may not be significant for the tissue investigated if the number of sample fields taken was too small.

To clarify this point, let us regard the fields as primary samples and the points as subsamples. The over-all variance $S_0{}^2$ is, then, given by

$$S_0^2 = \frac{S_p^2}{n} + \frac{S_s^2}{nm} \tag{3.29}$$

where S_p^2 is the variance among the primary samples and S_s^2 the variance of the subsamples within the primary units. From this formula it is evident that an increase in the number n of fields can reduce both partial variances, while an increase of the number m of points on the fields will influence only the second component.

With proper choice of magnification, 25 points per field appear, therefore, to be efficient and convenient for counting and promise a significant sampling, since, for example, 1000 points can be distributed over 40 fields. If desirable, the counts of four consecutive fields can be pooled and regarded as one sample, thus yielding percentages directly.

Due to the large variations in size of the different structural components it will be necessary to execute the volumetric analysis in a series of steps at different levels of magnification (see p. 52). Point counting is applicable to any of these levels: At the macroscopic level a sheet of celluloid or glass, on which a hexagonal lattice of (25) points of appropriate dimensions is drawn (Fig. 12 b), is superimposed on the cut surface of the specimen. For work at magnifications of a light microscope the Zeiss Integrating Eyepiece I, designed according to Hennig's recommendation, is convenient. Volumetric analysis at the electron microscopic level can be carried out directly on the fluorescent screen, if an appropriate system of points is engraved into the phosphor. If volumetry is the only measurement desired, a 25-point system is again the most appropriate. More possibilities are offered by use of the multipurpose screen described below which contains 30 points for volumetry.

At either level the differential counts can best be recorded by means of a mechanical differential counter, as used in hematology. These counters give a signal when 100 counts have been obtained but allow continuation if desired.

The total number of points necessary to estimate the volumetric proportion of the components under investigation within the required confidence limits will depend on the volumetric proportion ϱ itself. The probable relative error, E_ϱ, which is to be expected if n points have been counted is

$$E_\varrho = 0.6745 \cdot \sqrt{\frac{1-\varrho}{n \cdot \varrho}} . \tag{3.30}$$

In Fig. 13, E_ϱ has been plotted against n parametrically for various values of ϱ. It is evident that 1000 points are sufficient to keep the probable relative error below 5% in analyzing components which occupy only about 20% of the volume; 1000 points are rather rapidly counted. HENNIG has carried out a series of model test experiments and has found that the average value of ϱ obtained by point counting was always well within the probable error predicted by equation (3.30). In order to determine the number of points required for a desired accuracy, a rough preliminary estimate of ϱ is obtained by counting, for example, 100 points. The necessary number of counts can then be read from Fig. 13. Zeiss also supplies a nomogram for the determination of n along with Integrating Eyepiece I. If random sampling procedures are used in the preparation of the tissue for quantitation (paragraph III, A 2) the number n of points determined from Fig. 13 or from the nomogram should be counted on each of the primary samples, in order to obtain significant over-all results.

The number of counts necessary increases also with the degree of inhomogeneity of distribution of the component, which will mainly influence the number

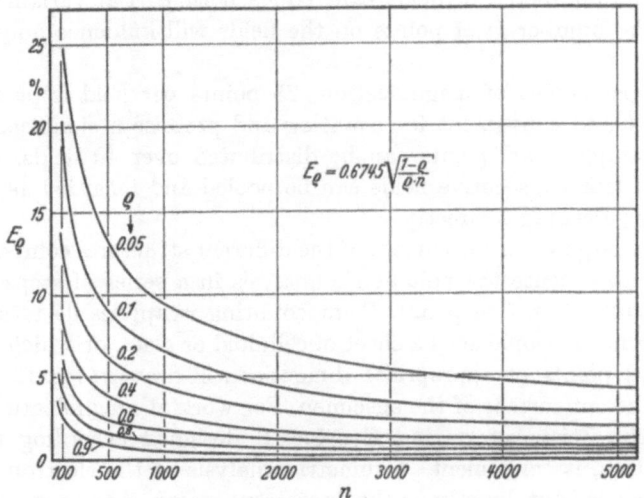

Fig. 13. Relative error E_ϱ for different magnitudes of the volumetric fraction ϱ as a function of the number n of sampling points investigated [From WEIBEL, 1963]

of sample fields required for significant sampling. Part of this inconvenience can be compensated by proper choice of a lower magnification.

The statistical analysis of the data obtained by point counting is usually restricted to a calculation of arithmetical mean and standard deviation. We prefer to obtain the latter by regarding a sequence of four fields of 25 points as one sample; however, each field may be taken as one sample, mainly if a large component (e. g. >60%) is analyzed. The half-width of the 95% confidence interval d is given by

$$d = t \cdot \sqrt{\frac{\varrho(1-\varrho)}{n}} \tag{3.31}$$

where t is the normal deviate for specified confidence limits (see p. 15), ϱ is the volumetric fraction determined, and n is the number of sample points.

c) Evaluation of the Point-counting and Linear Integration Methods

The point-counting method is simple and fast, and lends itself immediately to a wide variety of applications. The integrating eyepiece of Zeiss used in microscopic work is inexpensive; templates for macroscopic work are easily prepared, as are fluorescent screen lattices for electron microscopy. Point-counting may, however, not be as accurate as the linear integration methods, which are laborious and involve the use of considerably more expensive apparatus. HENNIG (1959) discussed these methods extensively and arrived at the conclusion that the point-counting methods are to be preferred for general application: Their rapidity allows drawing of larger samples, and mathematical supervision of the errors inherent in the method is easily possible.

Although this point of view is correct, the linear integration method using the Integrating Eyepiece of Leitz (SCHUCHARDT, 1954) may still have ist special place when several tissue components are to be analyzed volumetrically, mainly if some of the components are comparatively rare but evenly and finely distributed. This was, for example, believed to be the case for the respiratory portion of the lung with respect to the "tissue" component (see p. 54), in which instance it was preferred for this study. In retrospect, it is recognized that point counting would have had advantages as well, besides being more efficient. However, there is no reason to believe that the results could be significantly improved by repeating the same analysis by point counting.

d) Correction of Errors Introduced by Finite Section Thickness

The principle of DELESSE with its practical applications in the linear integration and point counting methods holds true only on infinitely thin sections or on cut surfaces observed by incident light. Microscopic sections, however, are slices of finite thickness which are observed by transmitted light. All opaque structures contained within the section space are thus projected onto the plane of. view, as illustrated in Fig. 14a. For correct application of the above principles we should, however, consider only those structures lying in the plane of the upper (or lower) cut surface. The projection of deeper structures will lead to an over-estimation of the volumetric proportions of prominent (opaque or dark) structures and, concurrently, to an underestimation of the surrounding, less conspicuous structures. In petrography this is known as the HOLMES effect (CHAYES, 1956). The degree of error introduced by this effect will depend on the relation between section thickness and dimensions of the structures under investigation. If the unit of the component is relatively large as compared to the section

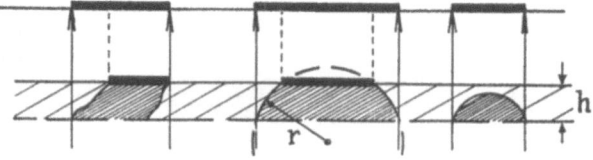

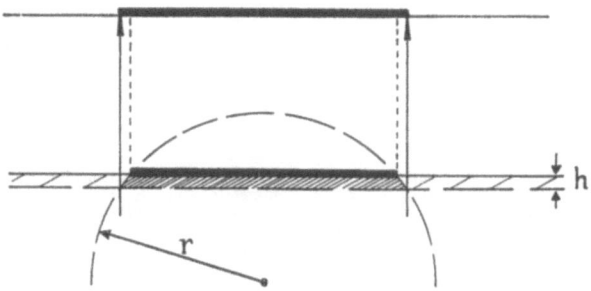

Fig. 14. Schematic illustration of HOLMES effect. Finite section thickness h causes opaque bodies to appear larger in projection than their actual transsection. The Holmes effect is less marked, if the size of the bodies is large as compared to the section thickness [From WEIBEL, 1963]

thickness, this error may be negligible in view of other errors and uncertainties inherent in the methods (Fig. 14b). If, however, the component is "fine granular", this error may be considerable. In practice, the dimensions of cells and other

components are comparable to the thickness of paraffin sections used in light microscopy, while larger tissue structures such as alveoli can be considered very large. In
electron microscopy, with section thicknesses of 600 Å, structures like pinocytotic
vesicles are "fine granular" with respect to the section thickness, but nuclei and
mitochondria are relatively very large.

The correction of the errors thus introduced during volumetric analysis will
depend mainly on the shape of the structures investigated. The correction factor
for approximately spherical, opaque bodies of radius r is well known (HOLMES,
1927; HENNIG, 1957). The true volumetric fraction ϱ' of these bodies is obtained
from the fraction ϱ determined on a section of thickness h through the relation:

$$\varrho' = \varrho \cdot \frac{4r}{4r + 3h} . \tag{3.32}$$

It can be seen that if h is very small, the error becomes negligible.

Equation (3.32) can be applied even if the bodies are only very roughly comparable to spheres; it will still improve the result.

In the volumetric analysis of the lung one of the components at the light
microscopic level ("tissue") required correction for section thickness. The interalveolar septa are the main con-
stituents of this component.
They are approximately spherical
shells of radius of curvature
$r \approx 100\ \mu$ and a thickness of 10μ.
Since the septum consists of the
shells of two adjacent spheres, the
actual thickness of the shell for
analytic purposes is $t \approx 5\ \mu$. With
a section thickness $h \approx 7\ \mu$ the
error introduced in estimating the
tissue fraction τ is not negligible.
The diagram of Fig. 15 shows, as
a model of this situation, a nor-
malized spherical shell of radius
of curvature 1 and a relative wall
thickness $\vartheta = \dfrac{t}{r}$. A section of relative thickness $\omega = \dfrac{h}{r}$ cuts out a
ring at position a or a cap at position b. In either case the area of
the opaque ring S projected by

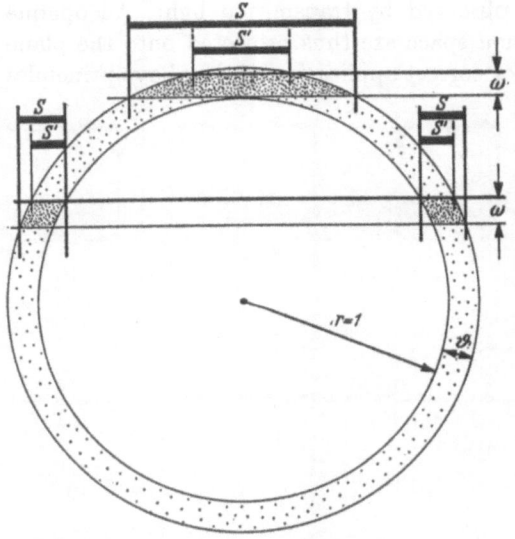

Fig. 15. Derivation of correction factor for overestimation
of "tissue"component τ in volumetric analysis of lung
parenchyma. It depends on the thickness of the interalveolar
septa ϑ and on the section thickness ω
[From WEIBEL, 1963]

parallel light rays is larger than its true area S' at either of the cut surfaces. They
are related as

$$S' = S \cdot f(\vartheta, \omega) . \tag{3.33}$$

The correction factor $f(\vartheta, \omega)$ is a function of the position of the slice with respect
to the center of the sphere. Calculation of an average coefficient $\bar{f}(\vartheta, \omega)$ will,
therefore, allow an appropriate correction of the data obtained on a large number

of such sections. The equation for $\bar{f}(\vartheta, \omega)$ applicable to all values of relative wall and section thickness is rather complicated and of little practical use in this context. If, however, $\omega \ll 1$ and $\vartheta \ll 1$ so that second order terms in these parameters can be disregarded, we find

$$\bar{f}(\vartheta, \omega) = \frac{\vartheta}{\omega} \log \frac{\vartheta + \omega}{\vartheta} + \frac{\vartheta}{1 + \vartheta} . \tag{3.34}$$

In application to this study the coefficient $\bar{f}(\vartheta, \omega)$ is found to be of the order of 0.66.

Through the principle of DELESSE we derive the volumetric fraction of a component from the fraction of the section area occupied by its projected areas of the shell section S. The correct tissue fraction τ^*, however, should have been obtained from the cut surfaces S'. We therefore find the correct value τ^* from

$$\tau^* = \tau \cdot \bar{f}(\vartheta, \omega) . \tag{3.35}$$

If the relative volume τ of the tissue shell of the alveoli was overestimated, the relative volume ϱ_A of the alveoli must have been underestimated. The correct value ϱ_A^* follows from

$$\varrho_A^* = \varrho_A + (1 - \bar{f}) \tau . \tag{3.36}$$

This correction will, of course, be small.

2. Principle for Counting Tissue Structures on Random Sections

The number of certain structures such as alveoli or renal glomeruli contained in a tissue volume is usually estimated by comparing the average dimensions of the single structures with the over-all volume they occupy (WILLSON, 1922; ELZE and HENNIG, 1956; ELIAS and HENNIG, 1961; and others). In the context of this study it appeared desirable to find a principle which would allow direct estimation of the number of structures from simple counts on histologic sections of the tissue (WEIBEL and GOMEZ, 1962).

a) Development of the Principle

The principle of DELESSE demonstrated that structures dispersed in a volume are quantitatively represented on sections of this volume. It can, therefore, be anticipated that the number of transsections of given structures found on the unit area of a random section bears some relation to the number of structures contained in the unit volume. It will be demonstrated that this relation depends on the volumetric density and on the shape of the structures (Fig. 16).

Suppose first N "granules" of equal size and shape with an individual volume v; they are randomly suspended in a volume V, occupying a fraction ϱ. From Delesse's principle it follows that

$$\varrho V = N \cdot v \tag{3.37}$$

$$\varrho S = n \cdot \bar{s} \tag{3.38}$$

where n is the number of transsections of the granules found in the area S of the section; $\bar{s}$ is the "mean cross-sectional area" of one granule obtained by sectioning it randomly in any direction (see below): it is related to the volume as

$$v = \beta \cdot \bar{s}^{3/2} \tag{3.39}$$

where β is a dimensionless coefficient depending solely on the configuration of the granules, as will be discussed later.

From equations (3.37), (3.38) and (3.39) we derive the following relationship between the number N of granules of equal size and shape contained in the unit volume and the number n of transsections of those granules on the unit area of a cut surface of the tissue:

$$N = \frac{n^{3/2}}{\beta \sqrt{\varrho}} \, . \tag{3.40}$$

Consider now an aggregate of N granules of equal shape but different dimensions. They are grouped into classes so that within each class i the individual

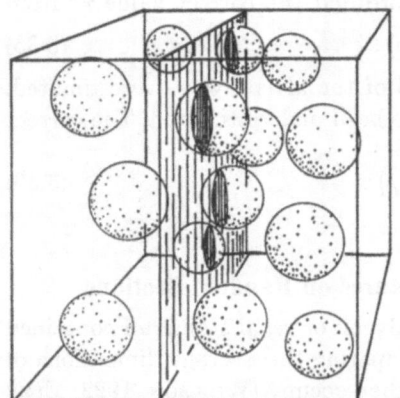 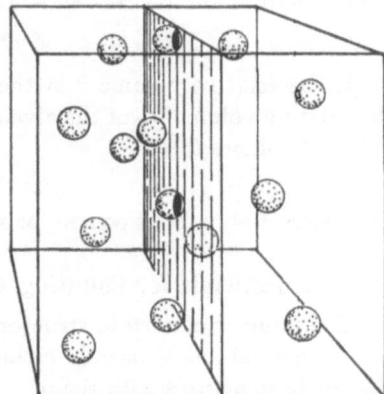

Fig. 16. Influence of volumetric fraction ϱ on the probability of sectioning n spheres out of N
[From WEIBEL, 1963]

volume $v_i = D_i^3$ is constant. Within each class, which has the numerical weight ξ_i, the relation

$$N_i = \xi_i N = \frac{n_i^{3/2}}{\beta \cdot \varrho_i^{1/2}} \tag{3.41}$$

holds, as demonstrated above. It now remains to find N in terms of the total number n of transsections of all granules on the unit area of a cut surface, and of the volumetric fraction ϱ of the containing volume occupied by the entire aggregate. Introducing appropriate weights for n_i and ϱ_i we obtain

$$N = \sum_i N_i = \frac{n^{3/2}}{\beta \cdot \varrho^{1/2}} \left[\frac{(\sum_i \xi_i \cdot D_i^3)^{1/3}}{\sum_i \xi_i \cdot D_i} \right]^{3/2} \tag{3.42}$$

$$\sum_i \xi_i D_i = (D)_1 \tag{3.43}$$

is the arithmetic mean, and

$$(\sum_i \xi_i D_i^3)^{1/3} = (D)_3 \tag{3.44}$$

is the cube root mean cube of the characteristic linear dimension D. $(D)_1$ and $(D)_3$ measure the first and third moment of the distribution. We thus obtain

$$N = \frac{n^{3/2}}{\beta \cdot \varrho^{1/2}} \cdot \left[\frac{(D)_3}{(D)_1} \right]^{3/2} . \tag{3.45}$$

In our first communication on this principle (WEIBEL and GOMEZ, 1962a) we have neglected[1] the distribution coefficient

$$K = \left[\frac{(D)_3}{(D)_1}\right]^{3/2}.$$

This would cause an underestimation of N, since $(D)_3 > (D)_1$. However, for "reasonable" size distributions of the granules K does not deviate much from 1. Calculating K for the distribution of glomerular diameters published by ELIAS et al. (1961) it was found to be 1.014; the error of 1.4% committed by neglecting K is evidently insignificant. For a symmetric distribution with a coefficient of variation of 25% $K = 1.07$. If the size distribution is expected to be very inhomogeneous $(D)_3$ and $(D)_1$ should be estimated, so that an appropriate K-value can be introduced. Very frequently, however, the error committed by neglecting K should be smaller than 5 to 10%, and it may be sufficient to use an arbitrary $K \approx 1.05$ in order to avoid a systematic underestimation of N.

b) The Coefficient of Configuration β

Equation (3.39) defined β as relating the average cross-sectional area of the granules to their volume. The determination of β for some typical bodies was presented in detail elsewhere (WEIBEL and GOMEZ, 1962a). It is comparatively easily done for bodies with an axis of rotational symmetry like cylinders or ellipsoids. In these cases the shape can be defined by the ratio ε of rotated axis to axis of rotation. For cylinders $\varepsilon_C = D/L$ and for ellipsoids $\varepsilon_E = b/a$, where a is the semiaxis of the rotated ellipse coinciding with the axis of rotation, and b is the semiaxis rotated about a. From this we find the coefficients β_C for cylinders and β_E for ellipsoids to be

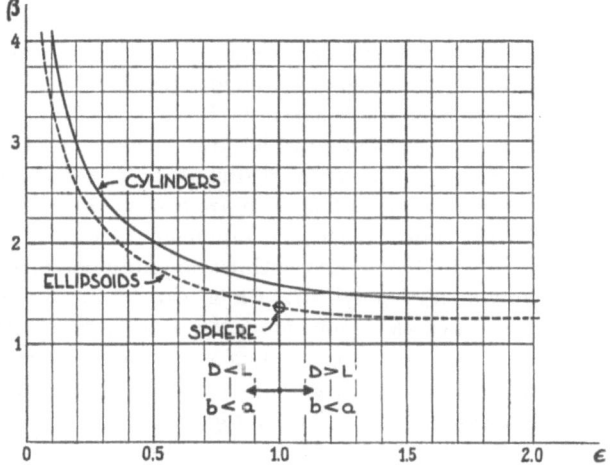

Fig. 17. Relation of the shape coefficient β of cylinders and ellipsoids to the ratio ε of shorter to longer axis (from WEIBEL and GOMEZ, 1962a)

$$\beta_C = \frac{2\pi}{\varepsilon \cdot \Phi(\varepsilon)^{3/2}} \quad (3.46)$$

$$\beta_E = \frac{4\pi}{3\varepsilon \cdot \Psi(\varepsilon)^{3/2}} \cdot (3.47)$$

The functions $\Phi(\varepsilon)$ and $\Psi(\varepsilon)$ are related to the probability of obtaining a given relative cross-sectional area by random sectioning of the granules (WEIBEL and GOMEZ, 1962a). The relations (3.46) and (3.47) are plotted in Fig. 17.

Coefficient β for other shapes. The sphere is obviously an ellipsoid with $\varepsilon = 1$. Its average cross-sectional area is well defined as

$$\bar{s} = (2/3) \cdot \pi r^2 \qquad (3.48)$$

[1] The author is grateful to Dr. R. DEHOFF for pointing out this omission, and to Mr. BRUCE W. KNIGHT for deriving the coefficient.

and therefore $\beta_s = 1.382 \ldots$ The value of β for hemispheres is similarly found to be

$$\beta_H = \frac{\sqrt{3} \cdot \pi}{4} = 1.36 .$$

Polyhedra which can be circumscribed to a sphere or to an ellipsoid of given ε will have a value of β slightly higher than that of the corresponding body with smooth surface since the corners will tend to reduce the value of $\bar{s}$. The higher the order of the polyhedron the closer to sphere or ellipsoid it will be. In general the β of polyhedra of the order of about 12 to 20 will be somewhere between ellipsoid and cylinder of corresponding ε. For prisms β will be somewhat higher than that of corresponding cylinders.

c) Application of the Principle

As reported elsewhere in detail (WEIBEL and GOMEZ, 1962a) the practical applicability of this principle was tested in a series of model experiments. These demonstrated a satisfactory degree of reliability with average errors of the order of 2% to 5%.

In practice this principle can be applied to the counting of structures dispersed in a tissue if the following conditions are satisfied:

1. The investigated bodies must be randomly distributed in the volume (see above).

2. They should be comparable to finite bodies, which are well defined in every direction, so that a coefficient β can be determined.

3. Typical size or shape groups must be randomly distributed (see above) and the largest possible size must be small as compared to the total containing volume.

4. The section on which transsections are to be counted should be very thin as compared to the smallest dimension of the investigated body.

This principle has been applied to the counting of alveoli and alveolar ducts. For this purpose these components of the respiratory zone could be regarded as randomly distributed within the pulmonary parenchyma (see p. 10). Alveoli have the shape of polyhedra which can be approximately circumscribed to a sphere. They have also some similarity to cylinders with a ratio $\varepsilon = 1$, or to a truncated cone. On the basis of such reasoning a value of the coefficient $\beta_A = 1.55$ was considered appropriate to represent alveoli. The coefficients of density ϱ_A were determined by the linear integration method described above. The average number of alveolar transsections was counted microscopically in each lung on rectangular fields measuring $9.1 \cdot 10^3$ cm^2 on the slide. These fields enclosed 25 to 50 transsections depending on the size of the lungs. Following conventional criteria, transsections which were entirely within the field were counted as well as those intercepted by the left and upper border line, while those intercepted by right and lower lines were neglected. The distribution of alveolar diameters can be supposed to have a coefficient of variation of $<25\%$ (compare STOREY and STAUB, 1962); the distribution coefficient K was therefore neglected (cf. p. 37).

Alveolar ducts and sacs are cylindrical structures of varying diameter-to-length ratio. The effective coefficient was therefore estimated to be the harmonic mean of the values of β pertaining to individual size groups. This could only be determined in an approximate manner, since the frequencies of the varying ratios were only roughly known. It was however found that a value of $\beta_D = 2$ would represent an appropriate estimate. The values of ϱ_D were again determined by linear integration (cf. p. 54). The number of duct transsections was counted in histological sections on fields measuring $5.9 \cdot 10^{-2}$ cm²; they included about 20 to 40 ducts. On each of five randomly selected sections five fields were sub-sampled and counted.

3. Methods of Microscopic Measurement of Linear Dimensions

Diameters and lengths were estimated with a conventional eye-piece micro-meter, in which the reticle is included in the plane of the intermediate image and can, therefore, be superimposed with a histologic section at any objective magni-fication. With proper adjustment errors arising from parallactic displacement of image and reticle should be insignificant. The eye-piece reticle was calibrated against a stage-micrometer of WILD; in various positions the calibrations showed good agreement.

Diameter and length of alveolar capillaries were measured on high power camera lucida drawings, mainly because the eye-piece micrometer did not provide enough precision. The scale of the drawings was again calibrated with a stage-micrometer. Since a high quality optical system was used (Wild Fluotar HI 50 oil immersion fluorite objective with compensating eyepiece) remarkably little distortion was apparent. There was no difference in dimensions noticeable if the same structure was drawn in the marginal or central portion of the projection field.

Measurements obtained on random sections of the structures under investiga-tion never correspond directly to some characteristic dimension of their shape. If, for example, a set of spheres of equal radius r_0 is sectioned randomly we will obtain a distribution of radii r of the transsections with probability densities

$$P(r, r_0) = \frac{r}{r_0 \sqrt{r_0^2 - r^2}} \, . \tag{3.49}$$

In this simple case r_0 can be inferred from the average value $\bar{r}$ of the measured section radii:

$$r_0 = \frac{4}{\pi} \bar{r} \, . \tag{3.50}$$

However, this condition will rarely occur. In most cases we shall be dealing with a distribution of sizes of the bodies investigated. For spherical bodies the probability density of measuring a radius r will be given by an integral equation of the form:

$$G(r) = \int_0^\infty P(r, r_0) \cdot F(r_0) \, dr_0 \tag{3.51}$$

where $F(r_0)$ is the distribution of radii of the spheres to be found. Such systems have been treated by WICKSELL (1925) and LENZ (1956). These authors have suggested practical ways of estimating the parameters of $F(r_0)$ from the distribu-tion of sectional radii measured on random sections. LENZ has presented a parti-cularly simple graphic solution. WICKSELL (1927) has also analyzed the situation where a mixture of ellipsoids of various sizes is sectioned randomly.

4. Principles for Estimating the Thickness of a Tissue Sheet such as the Alveolo-capillary Air-blood Barrier[1]

The alveolo-capillary air-blood barrier is a sheet of tissue of varying thickness τ which is bounded by two surfaces: the external surface S_E faces the air; the internal surface S_I is in contact with blood. The dimensional properties of such a sheet are characterized by the distribution of thicknesses $P(\tau)$. For many purposes it would be sufficient to determine the arithmetical mean thickness $\bar{\tau}$. Considerations on diffusion of gases, however, will require knowledge of the harmonic mean thickness τ_h which is defined as

$$\frac{1}{\tau_h} = \int\limits_0^\infty \frac{1}{\tau} P(\tau) \cdot d\tau . \tag{3.52}$$

The following principles are to provide us with the means to estimate $\bar{\tau}$ and τ_h as well as the relation of external to internal surface S_E/S_I by random measurements in the electron microscope[1].

a) Principle for Estimating the Arithmetical Mean Thickness $\bar{\tau}$

The arithmetical mean thickness $\bar{\tau}$ is an estimator of the amount of tissue lying between unit areas of the external and internal surface (Fig. 18). We therefore can define $\bar{\tau}$ as the ratio of tissue volume to bounding surfaces:

$$\bar{\tau} = 2 \cdot \frac{v_T}{S_E + S_I} = 2 \cdot \bar{\tau}_{1/2} \tag{3.53}$$

where $\bar{\tau}_{1/2}$ is the mean half-thickness of the tissue sheet which estimates the average amount of tissue per unit surface area without discrimination between

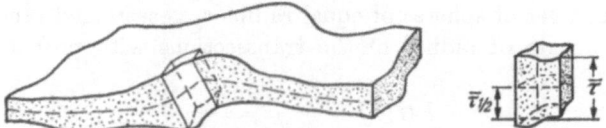

Fig. 18. Definition of arithmetic mean thickness $\bar{\tau}$ of air-blood barrier, and of mean half-thickness $\bar{\tau}_{1/2}$ in reference to bounding surfaces [From WEIBEL, 1963]

air and blood surface. For considerations on diffusion it may be more appropriate to relate τ to one or other of the bounding surfaces. The mean thickness $\bar{\tau}_c$ with respect to the capillary surface S_I would therefore be

$$\bar{\tau}_c = \frac{v_T}{S_I} . \tag{3.54}$$

Due to the negative over-all curvature of S_I, however, $\bar{\tau}_c$ will represent an overestimate of the mean air-blood distance as will be discussed.

CHALKLEY (1949) proposed a principle for estimating surface-to-volume ratios by means of randomly placed lines of finite length. A modification of Chalkley's method will allow us to derive $\bar{\tau}_{1/2}$ and $\bar{\tau}_c$ according to equations (3.53) and (3.54).

For the development of this principle, consider a piece of interalveolar septum contained in a cube of volume $V = L^3$ (Fig. 19). Now, for example, suppose a set

---

[1] These principles have been developed in collaboration with Mr. BRUCE W. KNIGHT, Affiliate at The Rockefeller Institute.

of N vertical lines passing through the cube and cutting through the horizontal planes with a density

$$\sigma = \frac{N}{L^2} . \tag{3.55}$$

The total length of these N lines is

$$\mathfrak{L} = N \cdot L \tag{3.56}$$

so that we can redefine

$$\sigma = \frac{\mathfrak{L}}{V} \tag{3.57}$$

as the density of lines in the volume V. These lines pass through the tissue and intersect its surfaces n times. At every point of intersection this line is forming

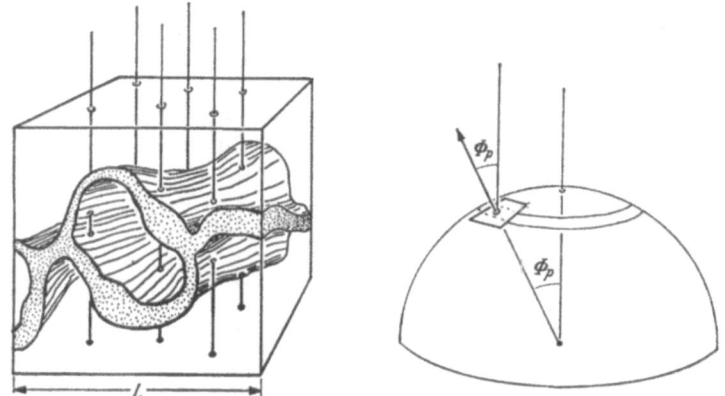

Fig. 19. Derivation of principle for estimating $\bar{\tau}$ and $\bar{\tau}_{1/2}$. Compare text [From WEIBEL, 1963]

an angle Φ_p to the normal of the surface (Fig. 19). If this tissue has a total surface area S we find that

$$n = \sigma \cdot S \cdot \langle \cos \Phi_p \rangle \tag{3.58}$$

where $\langle \cos \Phi_p \rangle$ is the average value of $\cos \Phi_p$. If this surface is randomly suspended in the cube V, it is equally likely for the normal to point in any direction. We therefore find the probability that Φ_p lies between Φ and $\Phi + d\Phi$ to be

$$P(\Phi + d\Phi > \Phi_p > \Phi) = \sin \Phi \, d\Phi \tag{3.59}$$

and, consequently

$$\langle \cos \Phi_p \rangle = \int_0^{\pi/2} (\cos \Phi) \, (\sin \Phi \, d\Phi) = \frac{1}{2} . \tag{3.60}$$

Equation (3.58) now becomes

$$n = \sigma \cdot S \cdot \frac{1}{2} = \frac{\mathfrak{L}}{2V} \cdot S \tag{3.61}$$

or, solving for S we obtain

$$S = \frac{n \cdot 2V}{\mathfrak{L}} \tag{3.62}$$

which corresponds, except for a factor of 2, to the equation of TOMKEIEFF (1945) and HENNIG (1956) presented on p. 37.

Suppose now in the same volume V a number P of points suspended randomly. A fraction

$$\varrho = \frac{p}{P} \tag{3.63}$$

of these points will lie within the tissue whose volume is

$$v_T = \varrho\, V = \frac{p}{P}\, V \tag{3.64}$$

as was shown on p. 19. By equation (3.53) we obtain the mean half-thickness of this tissue sheet as

$$\bar{\tau}_{1/2} = \frac{p}{P} \cdot \frac{\mathfrak{L}}{2n} \; . \tag{3.65}$$

We now break up the line of total length $\mathfrak{L}$ into Q lines of length z, so that

$$\mathfrak{L} = Qz \tag{3.66}$$

and distribute these lines randomly in space. They will have a total of $2Q$ endpoints which we can use as sampling points for volumetry if they are properly distributed. Putting therefore

$$P = 2Q \tag{3.67}$$

equation (3.65) becomes finally:

$$\bar{\tau}_{1/2} = \frac{z \cdot p}{4 \cdot n} \; . \tag{3.68}$$

In conclusion, equation (3.68) allows us to place a line of finite length z randomly in space a large number of times and to infer the half-thickness $\bar{\tau}_{1/2}$ of the given tissue sheet by counting the number p of endpoints lying in the tissue, and the number n of intersections of the line with its surface. The practical procedure used in applying this principle is outlined below.

From $\bar{\tau}_{1/2}$ the mean thickness $\bar{\tau}$ follows from (3.53). In order to obtain an estimate of $\bar{\tau}_c$ (equation 3.54) we should differentiate between the number n_I of intersections with the capillary surface and the number n_E of intersections with the alveolar surface. We then obtain

$$\bar{\tau}_c = \frac{z \cdot p}{4 \cdot n_I} \tag{3.69}$$

$$\bar{\tau} = \frac{z \cdot p}{2\,(n_I + n_E)} \; . \tag{3.70}$$

b) Relation of Alveolar to Capillary Surface

The differential count of the number of intersections of the sampling lines with internal (n_I) and external (n_E) surface of the tissue space allows us a direct estimate of the quantitative relation of the alveolar (S_E) to the capillary surface area (S_I). From equation (3.62) we obtain

$$\frac{S_E}{S_I} = \frac{n_E}{n_I} \; . \tag{3.71}$$

c) Principle for Estimating the Harmonic Mean Thickness τ_h

Consider first a flat sheet of tissue of thickness τ with two parallel boundaries (Fig. 20a). Place an infinitely long line randomly in space in such a fashion that it travels through the sheet over a length

$$l = \frac{\tau}{\cos \Theta} \qquad (3.72)$$

where Θ is the angle of inclination to the normal. Since Θ, measured in space, can assume any value from 0 to $\pi/2$, the length l will vary between τ and ∞. The distribution of l for given τ is therefore when $l \geqq \tau$

$$K(\tau, l) = \frac{\tau}{l^2} . \qquad (3.73)$$

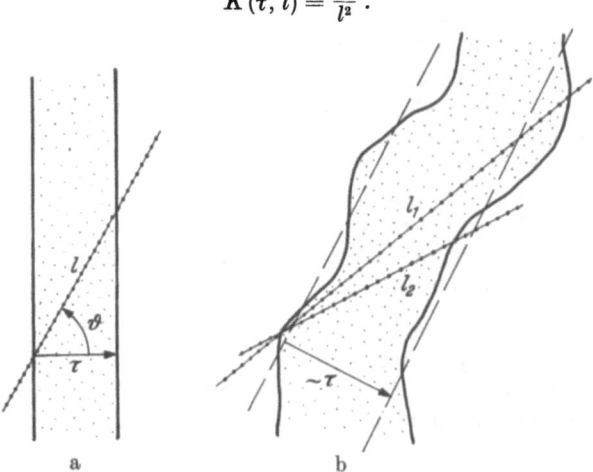

Fig. 20. Derivation of principle for estimating the harmonic mean thickness τ_h of air-blood barrier from length l of random line intercepting the tissue space. Compare text [From WEIBEL, 1963]

If we now take a large number of such sheets with a distribution of thicknesses $P(\tau)$ and place many lines at random across them, we will obtain a distribution of lengths

$$Q(l) = \int_0^l K(\tau, l) \cdot P(\tau) \cdot d\tau \qquad (3.74)$$

which is an integral equation. Given $Q(l)$ and $K(\tau, l)$ it is to find $P(\tau)$. The kernel $K(\tau, l)$ is given by equation (3.73) so that (3.74) becomes

$$Q(l) = \frac{1}{l^2} \int_0^l \tau \cdot P(\tau) \, d\tau . \qquad (3.75)$$

If $Q(l)$ is differentiable, we obtain

$$P(\tau) = \frac{1}{\tau} \frac{d}{d\tau} [\tau^2 Q(\tau)] . \qquad (3.76)$$

We are mainly interested in determining the harmonic mean thickness τ_h which is given by

$$\frac{1}{\tau_h} = \int_0^\infty \frac{1}{\tau} P(\tau) \, d\tau . \qquad (3.77)$$

By substituting (3.76) and integrating we obtain finally

$$\frac{1}{\tau_h} = 2 \int_0^\infty \frac{1}{l} Q(l) \, dl = 2 \frac{1}{l_h} \qquad (3.78)$$

or

$$\tau_h = \frac{1}{2} l_h \qquad (3.79)$$

which says that the harmonic mean thickness τ_h of the sheets is one-half of the harmonic mean length of the random measuring lines. If we obtain l in grouped measurements with classes denoted by G we estimate τ_h by:

$$\tau_h = \frac{1}{2} \cdot \frac{\sum\limits_{G=0}^{\infty} n_G}{\sum\limits_{G=0}^{\infty} \frac{1}{l_G} \cdot n_G} \cdot \qquad (3.80)$$

So far we have demanded that the sheets be flat, which in reality will not be the case at all. The alveolo-capillary tissue sheet will have numerous bends and will also be continuously changing in thickness (Fig. 20b). This will lead to an underestimation of τ_h which will depend on the degree of irregularity and on the curvatures of the sheet. Model experiments have shown, however, that τ_h is not underestimated by more than 20% so that the principle remains valid if appropriate precautions are taken (see below). An additional error may be introduced by the limitations of the field of view which does not allow us to measure very large values of l. Because we determine only τ_h, this error is negligible since the contribution of very long values of l is small.

d) Practical Application of Principles a — c

The air-blood tissue layer must be investigated on thin sections with the aid of an electron microscope. As discussed on p. 19, sampling points and sampling lines which are deposited randomly on a random section of tissue can be considered

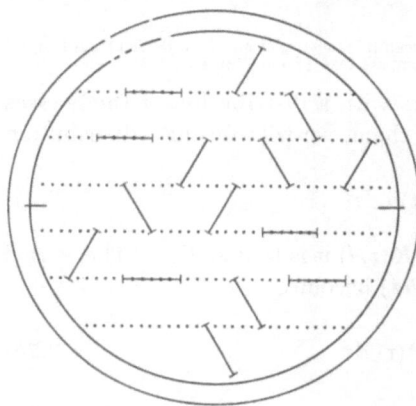

randomly distributed in space. We shall therefore devise a system of lines and points suitable for obtaining the measurements which are necessary for calculating $\bar{\tau}$ and τ_h, and shall incorporate it on the fluorescent screen of an electron microscope on which the tissue sections will be projected.

Fig. 21 shows the template of lines and points which will be suitable for this analysis. It contains a set of 15 lines of length $z = 1.2$ cm which are evenly distributed in three directions. The 30 end-points of these lines lie on the vertices of the equilateral triangles of a hexagonal point-lattice; their arrangement is thus well suited for point-counting volumetry (cf. p. 20). The distance of two such end-points is $z = 1.2$ cm, regardless of the direction; it is the same for points connected by a sampling line as for

Fig. 21. Lattice engraved on fluorescent screen of electron microscope for practical application of the principles derived in this section. Array of finite lines allows determination of $\bar{\tau}$ and $\bar{\tau}_{1,2}$. End points can be used for point count volumetry. Dotted lines are used for estimating length l of random intercept with barrier
[From WEIBEL, 1963]

points belonging to different lines. The presence of connecting lines therefore does not introduce a bias into point-counting volumetry. This system will allow the investigator to count the number p of end-points lying on sections of the tissue and the numbers n_I and n_E of intersections of the sampling lines with the internal and external surfaces of the tissue.

Six parallel rows of dots allow the determination of the lengths l of random lines passing through the tissue sheet by simply counting the number of dots lying on the tissue. The dots are spaced at equal distances $\triangle = 2$ mm. The end-points of the sampling lines subdivide the row of dots into groups of six, which is convenient for estimating long values of l. Conventionally, dots which touch the right-hand surface of the tissue are counted; those which touch the left-hand surface are rejected. Further details on the practical application of this system will be given on p. 100.

A fluorescent screen engraved with the pattern of Fig. 21 can serve various additional purposes. Simple point-counting volumetry can be done by obtaining differential counts of the 30 end-points of the line system; the rows of dots can serve as guide lines in obtaining this count. The inner circle has a diameter of 8 cm and encloses, therefore, a section area of 50.3 cm². It can be used for counting transsections if the number of certain structures is to be inferred by the counting principle of p. 25; transsections cut by the upper semicircle are counted; those cut by the lower are rejected. Using the equation proposed by CHALKLEY et al. (1949) the surface-to-volume ratio of any aggregate of bodies can be determined by means of the line pattern with end-points.

5. Principles for Estimating the Area of Three-dimensional Surfaces

In this context we are not so much interested in measuring the surfaces of individual structures as in estimating the collective surface of a set of unit struc tures building up an organ. As a model suppose a number of "granules" of wel-specified shape — say spheres — dis-persed randomly in a cube of given dimensions (Fig. 22): the total col-lective surface of all spheres in the cube is to be determined. In more specific application to the lung the units are represented by alveoli; the total alveolar surface to be found is defined as the sum of all indivi-dual alveolar surfaces.

There are basically two principles at our disposal which will allow the estimation of the collective internal surface of such subdivided organs. The first, called "geometrical", has the advantage of being descriptive of the geometry of the system. Its relations can, therefore, directly be used to correlate various param-

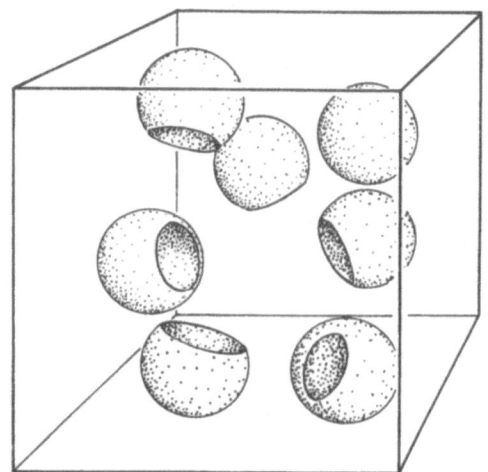

Fig. 22. Model for derivation of geometric method of estimation of alveolar surface area [From WEIBEL, 1963]

eters, e. g., over-all surface and volume (see p. 70). The second principle of "mean chord length" is very simple and reliable in application, and does not depend on any geometrical pattern of the system. It will therefore also be very useful in the study of pathological cases.

a) Geometric Principle for Surface Estimation

Consider a number N of well-defined bodies of equal size and shape contained in a cube of volume V (Fig. 22). Each body has a volume v and a surface s which are related as

$$s = \sigma \cdot v^{2/3} \tag{3.81}$$

where σ is a dimensionless coefficient which is independent of the size of the bodies but is fully defined by their shape. The collective relative volume ϱV and the collective surface S of these bodies is necessarily

$$N \cdot v = \varrho \cdot V \tag{3.82}$$

$$N \cdot s = S . \tag{3.83}$$

From these premises we obtain the relation

$$S = N \cdot \sigma \cdot \left(\frac{\varrho \cdot V}{N} \right)^{2/3} . \tag{3.84}$$

Suppose now that the volume V contains an aggregate of N bodies of equal shape but different dimensions. They are grouped into classes so that within each class i the individual volume $v_i = D_i^3$ is constant. Each class contains $N_i = \xi_i \cdot N$ bodies, occupies a fraction ϱ_i of the total containing volume, and, according to (3.84), has a collective surface

$$S_i = N_i^{1/3} \cdot \sigma (\varrho_i V)^{2/3} . \tag{3.85}$$

The coefficient σ is the same for all sets since all bodies have an identical shape. The over-all collective surface of the entire aggregate is then

$$S = \sum_i S_i .$$

Introducing ξ_i as the weighting factor for N_i and an appropriate weighting factor for ϱ_i we obtain

$$S = N^{1/3} \cdot \sigma (\varrho V)^{2/3} \cdot \left[\frac{\left(\sum_i \xi_i \cdot D_i^2 \right)^{1/2}}{\left(\sum_i \xi_i \cdot D_i^3 \right)^{1/3}} \right]^2 \tag{3.86}$$

Equation (3.86) allows us to estimate the internal surface S of the aggregate of bodies, if we know their total number N and the volumetric fraction ϱ of the containing volume occupied by them. These quantities can be determined on sections by principles outlined above.

$$\left(\sum_i \xi_i D_i^3 \right)^{1/3} = (D)_3 \tag{3.87}$$

is the cube root mean cube of the variable D_i, and

$$\left(\sum_i \xi_i D_i^2 \right)^{1/2} = (D)_2 \tag{3.88}$$

is its root mean square. The ratio

$$G = \left[\frac{(D)_2}{(D)_3} \right]^2 \leq 1 \tag{3.89}$$

is thus a measure of the spread of the distribution which is again not very sensitive [cf. equation (3.45)]. For a somewhat symmetric distribution with a coefficient of variation of 25% $G = 0.96$. For the distribution of glomerular diameters published by ELIAS et al. (1961) $G = 0.99$. Neglecting the distribution coefficient G would thus have resulted in a slight overestimation of S. In many instances, however, G will be very nearly equal to 1, so that neglecting it would result in an inappreciable error.

If the value of N has been obtained by the principle of counting transsections of the bodies outlined on p. 25 the distribution factor influencing the final relationship is

$$F = K^{1/3} \cdot G = \frac{(D)_2^2}{(D)_1^{1/2} \cdot (D)_3^{3/2}} \cdot \tag{3.90}$$

This coefficient is very insensitive to the spread of the distribution. For the above example of a coefficient of variation of 25% $F = 0.993$. In determining parameters like the alveolar surface area of normal lungs it may thus be well permissible to neglect the distribution factor and to estimate S from

$$S = \tilde{N}^{1/3} \cdot \sigma \cdot (\varrho V)^{2/3} \cdot \tag{3.91}$$

Here $\tilde{N}$ is the number of bodies obtained from equation (3.45) by disregarding the distribution coefficient K. Equation (3.91) corresponds thus to (3.84).

The coefficient of configuration σ can be calculated from equation (3.81) if the shape of the bodies can be assimilated to a geometrical figure whose volume and surface can be calculated. Figure 23 presents the values of σ for a series of typical shapes: they are bodies of rotational symmetry whose geometrical properties are expressed in terms of the ratio ε of the rotated semiaxis to the semiaxis of rotation as defined on p. 27. If in an aggregate the shape of the bodies varies independently of the variation in size the value to be entered into equations (3.86) or (3.91) is the arithmetic mean of σ. If the variation of σ depends on the size of the bodies a weighted mean has to be introduced, the weighting factor being $\xi_i \left[\dfrac{D_i}{(D_3)} \right]^2$.

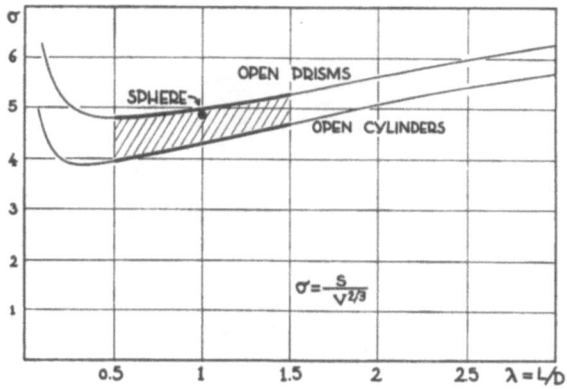

Fig. 23. Coefficient σ relating surface and volume of geometric bodies which show similarities to alveoli plotted as function of the ratio λ of length to diameter. Compare text [From WEIBEL, 1963]

b) The Principle of "Mean Chord Length"

A simple method for the determination of the "internal surface" of subdivided volumes has been proposed, apparently independently, by TOMKEIEFF (1945) and HENNIG (1956). It is based on the following principle:

If a line of known length L is thrown randomly on a random section of lung, it will intersect interalveolar septa m times (Fig. 24). If this is done a sufficient number N of times

$$\frac{N \cdot L}{\sum\limits_{i=1}^{N} m_i} = L_m \tag{3.92}$$

will represent the "mean chord length" of the randomly placed line with alveolar membranes. TOMKEIEFF and HENNIG have shown that the over-all surface S_{AT} of the double-faced alveolar membrane follows from the equation

$$S_{AT} = \frac{4 \cdot (\varphi \cdot V_L)}{L_m} = 4 \cdot \frac{\sum m}{N \cdot L} \cdot (\varphi V_L) \qquad (3.93)$$

where (φV_L) is the total volume of the respiratory portion of the lung. We shall point out that this amounts to about 90% of the total lung volume V_L.

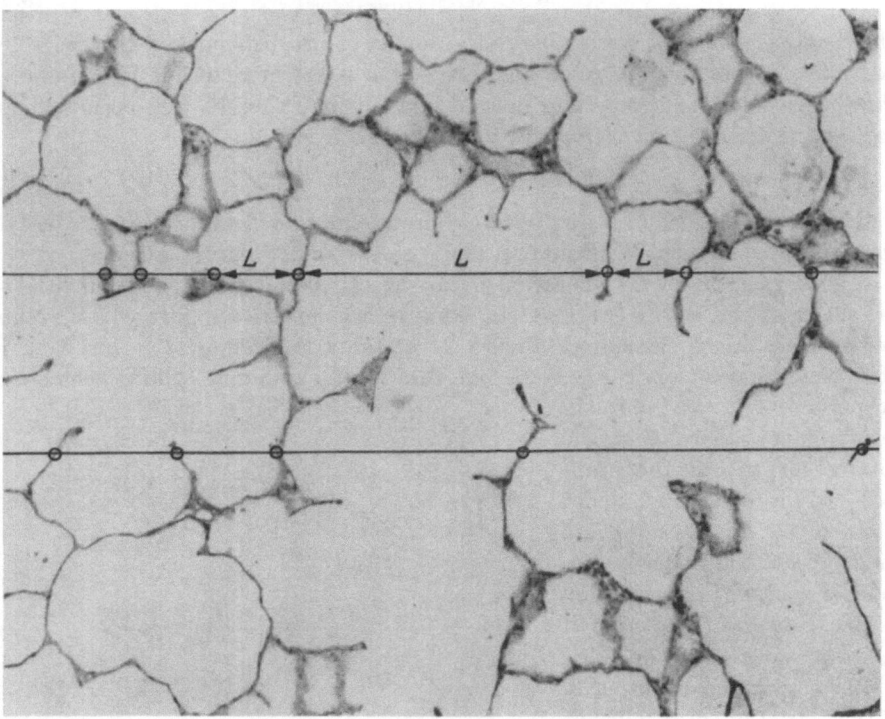

Fig. 24. Section of parenchyma of human lung with two parallel lines intersecting interalveolar septa at circles. "Mean chord length" is estimated as the average of the lengths L between intersections. Compare text

HENNIG has provided this principle with a solid mathematical basis. It has corollaries in the studies of CROFTON (1898) on the "average length of chord" of a line intersecting with a plane figure in a two-dimensional system, and in the method for estimation of volume-surface ratios developed by CHALKLEY et al. (1949) (cf. p. 32).

We applied this principle to the determination of the total alveolar surface in our five lungs by proceeding in the following manner: A microscopic eye-piece reticle was provided with a square and a hairline cross. The lengths of the vertical and horizontal lines of the cross were calibrated with a stage micrometer; they were identical. Two lines vertical to each other were considered in order to compensate for unavoidable distortion of the paraffin sections because of compression due to sectioning. These measurements can also be carried out by using the Zeiss Integrating Eyepiece II designed by HENNIG.

This reticle was optically superimposed on histologic sections approximately 7 μ thick. The number m of intersections of the vertical and horizontal lines of the cross with alveolar membranes was counted. This was done 10 times on each of five sections selected at random from various regions of the lung (see p. 45); the yield was thus 100 measurements. The "mean linear intercept" L_m was then calculated according to equation (3.92), and the total alveolar surface S_{AT} followed from equation (3.93).

c) Volume-to-Surface Ratio (Coefficient σ)

CHALKLEY et al. (1949) proposed a method for estimating surface-to-volume ratios of bodies suspended in a volume which, basically, amounts to a combination of point-counting volumetry and surface determination by "mean chord length". If a line of known length is thrown on a random section the surface-to-volume ratio can be estimated from the number of "hits" of the bodies in consideration by the end-points of the line and from the number of intercepts of the line with their surface (cf. p. 32).

We have estimated the surface-to-volume relationship of the single alveolus by combining the two independent methods of surface estimation presented in the preceding sections. When one introduces the value of the alveolar surface obtained from equation (3.93) into equation (3.91), the only unknown is the shape coefficient $\bar{\sigma}_A$ for which we can solve to obtain the equation

$$\bar{\sigma}_A = \frac{4 \cdot (\varphi \cdot V_L)^{1/3}}{N_{AT}^{1/3} \cdot \varrho_A^{2/3} \cdot L_m} \cdot \tag{3.94}$$

Each of the five factors on the right side of this equation was determined independently. The factors involve: the total number of alveoli N_{AT}, their fraction ϱ_A of the volume (φV_L) of the respiratory portion of the lung and the "mean chord length" L_m.

Chapter IV

Material and Methods of Preparation for Morphometric Studies of the Lung

A. Material Used in this Study

The arrangement of the elements of the respiratory and transitory zones satisfies the conditions for "random distribution" outlined on p. 10. On the other hand, the elements of the conductive zone show a distinct polar orientation with origin in the hilum, or, more properly, in the mediastinum; they are, therefore, non-randomly distributed. This fact necessitates the use of different sampling procedures for the quantitation of respiratory and transitory lung structures, than those used for the measurement of conductive elements, as described in Chapter III. Accordingly, it is necessary to prepare the material by different methods. Peripheral lung structures, because of their small size, will be studied by microscopic techniques, while a large part of the conductive elements will be accessible to macroscopic investigation.

1. Material for Microscopic Studies

For this part of the study five normal human lungs of varying age and size were used (Table IV.1). They were obtained from autopsies on persons who died suddenly from various conditions unrelated to the lung, mostly from accidents. These lungs were carefully removed from the chest; one of the two lungs from each subject was at our disposition, while the other was reserved for routine pathologic examination. Prior to sampling, the five lungs used by us were selected from among a larger number of lungs prepared in the same manner as best satisfying the conditions of appearing normal to gross examination. "Normality" was mainly characterized by the absence of significant pathologic changes such as "emphysematous foci", signs of chronic or acute inflammation, or distortion of airways. A slight degree of pulmonary edema, present in two instances was considered not to interfere seriously with any of the information vital to this study. The five subjects selected represented different age groups varying between 8 and 74 years. This fact, of course, limits the feasibility of interpreting significant changes due to ageing. Such factors will need to be confirmed subsequently by study of a larger number of lungs in each age group.

Table IV.1. *Cases Investigated*

Age	Sex	Lung weight gm		Lung studied	Lung volumes ml		J ml/gm	ζ	Cause of death
		R	L		V_{LF}	V_L			
8	m	210	210	R	1350	2580	6.4	0.8	spontaneous cerebral hemorrhage
16	f	220	210	L	1400	2680	6.5	0.8	auto accident; ruptured spleen, severe hemorrhage
34	m	450	500	L	3000	5750	6.0	0.75	subway accident; amputation of thighs, shock
48	m	600	600	R	3700	7100	5.9	0.74	traumatic skull fractures; subarachnoidal hemorrhage
74	f	500	500	R	3100	5950	6.0	0.75	multiple skull fractures

The over-all lung dimensions were defined by weight of the organ at autopsy and by its total volume. The weight was assumed to reflect the constitutional size of the lung and was determined separately for left and right half-lung (Table IV.1). The total volume of the inflated lung obviously depends on the constitutional lung size and on the degree of inflation (see p. 44). In the course of this study we shall make reference to different types of "lung volumes", defined as follows:

V_{LF}: *Volume of the fixed lung*. This baseline volume, on which all further calculations will be based, was measured by water displacement following full fixation of the half-lungs ($V_{L/2F}$). The volume of the entire lung was computed on the basis of the ratio of total lung weight ($W_R + W_L$) to the weight W_H of the half-lung which was used in the study:

$$V_{LF} = \frac{W_R + W_L}{W_H} \cdot V_{L/2F} .$$

(4.1)

It should be noted that this volume includes air, tissue and blood.

V_L: *Volume of the fresh lung before fixation*, again including air, blood and tissue. This volume is computed from V_{LF} on the basis of the correction factor f^{-3} (Table IV.2) which takes account of the shrinkage of the lung due to fixation and retained ability of recoiling:

$$V_L = f^{-3} \cdot V_{LF} . \tag{4.2}$$

V_{LS}: *Volume of the processed lung*, referring to the state of the lung tissue as it is observed on histologic preparations, after it has passed through the standardized process of dehydration, paraffin embedding and sectioning (p. 45). It is obtained from V_{LF} by multiplication with the correction factor p^3 (Table IV.2):

$$V_{LS} = p^3 \cdot V_{LF} . \tag{4.3}$$

All dimensions obtained on histologic preparations refer to this volume.

TLC: Total lung capacity, which is the amount of gas contained in the lung at the end of a maximal inspiration (cf. COMROE et al., 1955). *TLC* is a physiologic measurement in the living which is not necessarily related to the anatomic lung volumes used here since it may be influenced by impaired expansion of the chest or mobility of the diaphragm. In the normal it must, however, be somehow related to the maximal degree of inflation which can be obtained in vitro on the eviscerated cadaver lung.

V_M or V_{TLC}: *Anatomic lung volume at the degree of inflation corresponding to TLC*. In addition to the air volume *TLC*, this includes the volume of pulmonary tissue and blood, which is of the order of magnitude of 12% of *TLC*, so that

$$V_{TLC} \approx 1.12 \, (TLC) .$$

In an attempt to normalize the degree of inflation of the lungs investigated, their volumes V_L will later be expressed as fractions ζ of V_{TLC} (cf. p. 44, and Table IV.1).

2. Material for macroscopic Study of Conductive Airways

The dimensions of major conductive airways were studied on plastic casts of the human bronchial tree, which were prepared and kindly lent to us by Dr. AVERILL A. LIEBOW of Yale University. One of these casts was selected for the study of the dimensional changes of bronchi progressively distant from the airway origin. The over-all dimensions of this specimen were estimated to compare well with those of an average adult human lung as will be discussed later (see p. 116). Since the preparation of such casts introduces many uncertainties, some dimensions were also obtained from bronchograms of the living, which were prepared and kindly loaned to us by Dr. THANE ASCH of New York Hospital and Cornell University Medical School. These data were used for comparison with the measurements obtained from casts; it was also possible to relate lung size and bronchial dimensions on this basis (see p. 132).

B. Preparation of the Lungs for Microscopic Studies
1. Fixation and Processing of Lung Tissue

The most widely employed method utilizes the instillation of the fixative — mostly formaldehyde — into the airways (MOOLTEN, 1935; HARTROFT and MACKLIN, 1943; LOOSLI and POTTER, 1959). However, this procedure drastically alters one of the essential "natural" conditions of the lung in that it substitutes fluid for air as the content of the airways. Among the many changes that may occur, it is well known that the apparently important forces generated by surface tension at the air-liquid interphase of the alveolar surface are abolished (VAN NEERGARD, 1929; RADFORD, 1957; and others). In addition, the heavy fluid mass will cause distortions of the architecture of the delicate peripheral structures. Therefore, it would appear essential to employ a more "natural" fixation technique.

In recent years, several methods have been developed, which allow fixation

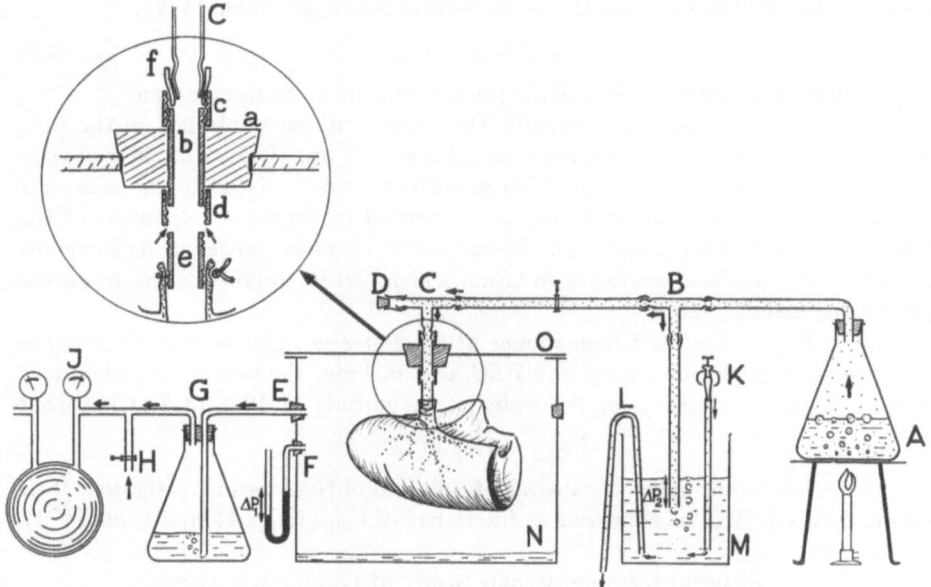

Fig. 25. Apparatus for fixation of lungs with formalin steam at controlled degree of inflation. *A* flask with formalin solution; *B* water escape valve for pressure control in steam line; *C* connector to main bronchus consisting of threaded brass tubing (*a—e*) in rubber stopper (*a*); sleeve of soft rubber hose (*f*) adapts *T*-tube *C*; *D* side arm for air inflation; *E—J* pumping system for producing controlled negative pressure in chamber *N* (from WEIBEL and VIDONE, 1961)

of the lung in its expanded state by means of compressed air (TOBIN, 1952) or formaldehyde vapors or fumes (BLUMENTHAL and BOREN, 1959; PRATT and KLUGH, 1961). Besides being quite time-consuming (several days are required for fixation), these techniques produce a dry lung. The extraction of water from the delicate tissues of the lung without replacement by another fluid inevitably results in considerable shrinkage and distortion of the structures, the geometry and dimensions of which we wish to explore.

Because of the limitations inherent in these techniques, we have developed a method of lung fixation by means of concentrated formalin steam applied to the lung via the airways. A detailed technical report of this method has appeared elsewhere (WEIBEL and VIDONE, 1961).

The lungs used in this study were fixed by this procedure in a controlled degree of gaseous inflation. At autopsy both half-lungs were removed from the chest, carefully avoiding any damage to their surface. The half-lung which was placed at our disposal was mounted in a plastic box (Fig. 25) with the bronchus open to the outside. The lung was then inflated by negative pressure which was developed within the box by means of an air pump. A screw clamp on a side-arm (*H*) made it possible for the lung to be "breathed" by changing the resistance in that tube, thus permitting gradual inflation. The pressure gradient was controlled by means of the manometer *F* and maintained at around 10 cm of water with short intermittent increases up to 20 cm in order to gradually open all collapsed areas of the lung.

Formaldehyde steam was produced in flask *A* containing about 1000 ml of 37% formaldehyde and 500 ml of water. When the lung was fully inflated to the desired degree (see below) the steam was led over a water escape valve into the airways. Fixation of the lung tissue was completed in about 2 to 3 hours. After cooling, the lung was gently placed in a large pail containing Zenker's solution and covered with a layer of wet cotton. Overnight an external layer of lung tissue was, thus, hardened so that the specimen could be sectioned serially into slices of about 1.5 cm thickness by aid of 18″ Gillette Stady blades without appreciable deformation (Fig. 26). The slices were then loosely wrapped in gauze and reimmersed in Zenker's solution for 24

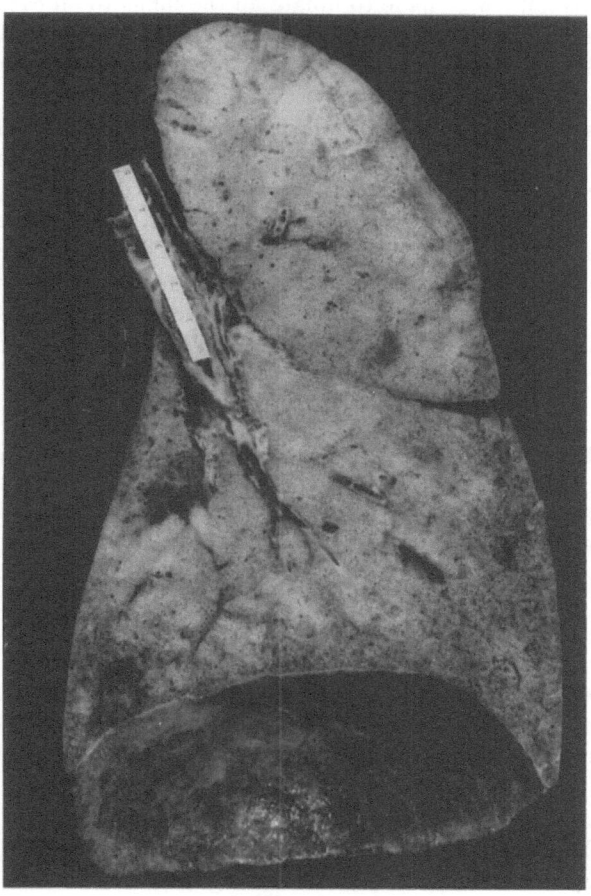

Fig. 26. Cut surface of young woman's lung fixed by formalin steam method. Lung was allowed to stay in Zenker's solution over night. Discolorations are due to uneven penetration of Zenker's fluid (from WEIBEL and VIDONE, 1961)

hours. By that time the lung tissue had become sufficiently firm to withstand manipulation, cutting of blocks, and embedding.

The tissue blocks (see below) were processed through alcohol, methyl benzoate, and xylol into paraffin. Negative pressure was applied for several hours at various stages to assure adequate penetration and removal of air bubbles. Among the various paraffins "Bioloid" was found to give the best results with lung tissue. Sections of 7 to 8 μ and 50 to 100 μ thickness were cut on an ordinary rotating microtome.

The preservation of tissue structure by this fixation procedure was excellent (WEIBEL and VIDONE, 1961).

2. Degree of Inflation

It is of prime importance to know at what degree of inflation the fixation of the lungs, on which measurements are to be carried out, has occurred. Ideal conditions would be provided by fixing the lungs in a plethysmograph. However, a variety of circumstances did not permit this for our material. Accordingly, an attempt was made to inflate all the lungs to an easily definable point along their pressure-volume curves (RADFORD, 1957; MEAD, 1961). When all areas of the lung had been aerated by "breathing" manipulations (see above), the course of the pressure-volume curve was estimated by inspection. It was usually rather easy to see when the point of inflection had just been passed on inflation. The lungs were then maintained at this degree of inflation, which is facilitated by the well-known static hysteresis of the pressure-volume relationship in inflation and deflation.

After refixation in Zenker's solution, the volume of the fixed lung was determined by water displacement (Table IV. 1). The volume of the fresh lung could be calculated on the basis of correction factors still to be discussed.

Secondarily, we attempted to estimate the relative degree of inflation J by correlating the volume of the fresh lung V_L with its autopsy weight W:

$$J = \frac{V_L}{W} . \tag{4.4}$$

If we assume as a baseline that the average adult human lung reaches a total lung capacity (TLC) of 6000 to 7000 cm³ and weighs, at autopsy, 900 to 1000 gm then the average maximal lung volume (air, tissue and blood) is approximately 6900 to 8000 cm³. The coefficient J_{TLC} relating volume to weight at total lung capacity is, therefore,

$$J_{TLC} = 8 \text{ cm}^3/\text{gm} .$$

The values of the inflation coefficients J for the five lungs used in this study are listed in Table IV.1 Their average is $\bar{J} = 6.2 \pm 0.26$. By relating the observed coefficient J to the baseline coefficient J_{TLC} we can determine an index of the relative degree of inflation:

$$\zeta = \frac{J}{J_{TLC}} . \tag{4.5}$$

These values are listed in Table IV.1; the average degree of inflation corresponded approximately to "³/₄ total lung capacity". Although this correlation is only approximate it, nevertheless, allows us to relate the subsequently presented dimensions to a physiologically rather well-defined functional state, even if only in orders of magnitude. The baseline values of J_{TLC} and of coefficients relating to other physiologic states remain to be more accurately determined.

The rather small variation of ζ in our material gives us confidence that the lungs had been fixed at approximately corresponding states of inflation, so that dimensions obtained on these lungs can be compared with each other validly.

3. Correction for Artifacts

During the entire preparation procedure described, the lung tissue suffers a considerable degree of shrinkage. This occurs during fixation as well as in dehydration for embedding. The use of hot formalin vapors for fixation may have caused a somewhat higher degree of formalin shrinkage than is usually expected from formalin immersion fixation. In order to allow conversion of the results of our measurements into the "fresh state" before fixation, we have estimated, stepwise, the factors by which the tissue dimensions are reduced by our procedures (Table IV.2).

Table IV.2. *Correction Factors for Artificial Shrinkage*

Conversion		Factor		
from	to	Linear	Area	Volume
fresh tissue	fixed tissue	$f = 0.82$	$f^2 = 0.67$	$f^3 = 0.55$
fixed tissue	fresh tissue	1.22	1.50	1.82
fixed tissue	processed tissue	$p = 0.92$	$p^2 = 0.84$	$p^3 = 0.78$
processed tissue	fixed tissue	1.09	1.18	1.28
fresh tissue	processed tissue	$f \cdot p = 0.74$	$(f \cdot p)^2 = 0.55$	$(f \cdot p)^3 = 0.41$
processed tissue	fresh tissue	1.35	1.82	2.46

a) The volume V_L of the fresh lung, inflated to the level of 3/4 TLC described, was estimated from orthographic tracings of the lung contours prior to fixation. The volume V_{LF} of the fixed lung was measured by water displacement. The ratio $\dfrac{V_{LF}}{V_L} = f^3$ yielded the "fixation shrinkage factor" for volumes which averaged 0.55. Square and linear factors were derived from this value. The linear factor f was also estimated directly by comparing characteristic linear dimensions of the orthogram with identical dimensions measured on slices of the fixed lung. All the values were averaged. The reciprocal factors allow conversion of "fixed dimensions" into "fresh dimensions".

b) The shrinkage factors p, due to paraffin embedding and sectioning, were estimated from the comparison of the linear dimensions of the tissue blocks prior to processing with the same dimensions measured on the stained sections, thus allowing conversion of "fixed dimensions" into "processed dimensions". Again, the values $\dfrac{1}{p}$ allowed inverse conversions.

c) Cumulation of the factors f and p yielded factors for the conversion of dimensions between the "processed" and the "fresh" state.

4. Practical Methods of Random Sampling

The tissue samples used for microscopic measurements were obtained by a two-stage sampling procedure that was defined in general terms on p. 13. It remains now to outline practical methods by which these samples can be obtained.

a) Primary Sampling

The first sampling stage had to provide a set of histologic preparations. This was again obtained in two substages. The fixed lung was cut into parallel slices 1.5 cm thick (see above). A set of 12 to 20 blocks was taken from various regions of these slices in an attempt to distribute the samples as evenly as possible over all parts of the lung. Thus, an element of stratification was introduced. From the collection of histologic preparations, obtained by sectioning these blocks, a sub-sample of five slides was drawn, again in the manner of stratified random sampling.

Dividing the total set of slides into two strata, upper and lower parts of the lung, it was observed that either of these two regions should be represented by two or three sections. From each stratum, however, the slides were drawn blindly.

This procedure was adopted at the start of the study, and all material discussed here was sampled in this fashion. In retrospect, however, it can be said that this method is not ideal if rigorous random sampling rules are to be followed. Since we were dealing with normal lungs, in which no gross irregularities should occur, we have decided that this approach was satisfactory for our purpose. This judgment is supported by the fact that the confidence requirements outlined above were always well satisfied. Moreover, the various results obtained on five lungs with quite different characteristics proved to be consistent.

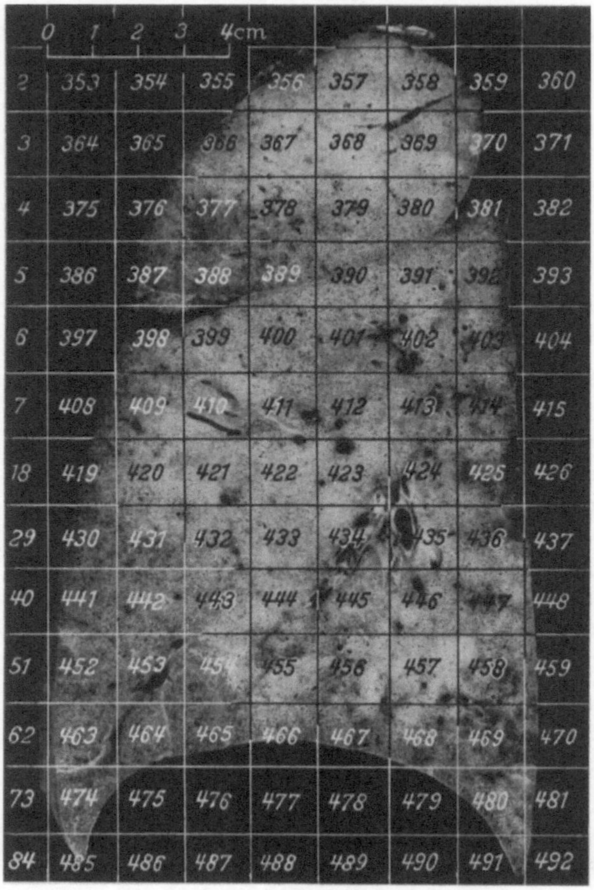

In the course of this study a more rigorous sampling procedure was developed which is both more efficient and more reliable. It is presented here as a suggestion for use in future studies, mainly on pathologic material.

In this version, the fixed lung is again cut into slices 1.5 cm thick, which are laid out on a table. On the upper cut surface of all slices a celluloid sheet, on which a

Fig. 27. Slice of same lung as in Fig. 26 with grid of squares of 1.5 cm diameter for random number sampling. Compare text

square lattice with 1.5 cm spacing is drawn, is superimposed at random (Fig. 27). As an alternative, the lattice can be chosen to be rectangular, with 1.5 by 2.5 cm spacing. Each field is numbered consecutively and continuously over all slices. If n units are required in the sample, n figures are picked from a table of random numbers following routine procedures. This fixes the location of the sample blocks, which is marked on the lung tissue by inserting headless pins through the perforated corners of the squares. The blocks are carefully cut out with razor blades, numbered 1, 2, 3, ..., n in the order in which their location was read off the random number table; they are then processed and sectioned. The slides thus obtained form the primary sample. It is advisable to take a few extra samples in excess of n, which are numbered $n + 1, n + 2, \ldots$ and can be used as replacements if any accidents should occur during processing. If there is doubt about the sample size n, a second sample can be taken which is numbered from $n + 1$ to $2n$. If the results obtained from the first sample should not satisfy the specified confidence requirements, supplementary sections can be drawn from this second sample.

It is evident that this procedure can readily be modified to allow stratified random sampling. This may, for example, be in the interest of significant sampling on pathologic specimens where the lesions are not diffuse but appear concentrated in some regions. It is likewise possible to use this general method in sampling other organs.

b) Subsampling

From each block a single section was used for the analysis. The area of the sections was much larger than the field which could be investigated microscopically at practical magnifications. It was, therefore, necessary to subsample the sections by a random procedure. This was achieved by blind random displacement of the mechanical stage of the microscope without observing through the tube. To eliminate bias, the horizontal and vertical stage micrometers were set each time at an integer millimeter position. Random numbers could also be used in selecting the sampling position, but this proved to be too cumbersome to be justified at this stage. If a large fraction of the section area is to be covered with sampling fields, it is advisable to use a systematic sampling procedure, in order to avoid overlap of the fields.

5. Methods of Histologic Demonstration

The peripheral *airway structures* of the lung could be investigated morphometrically in the light microscope at low magnification using histologic sections of about $7\,\mu$ thickness. These thin sections were stained with H & E, Goldner's trichrome, Weigert's elastica counterstained with pikroindigocarmine-Kernecht-rot, and with the PAS reaction.

The *alveolar capillaries* had to be demonstrated by a special technique. Forming a dense network within the interalveolar membrane (see below), they are enveloped by a fine continuous basement membrane which has the property of staining intensely positive with the PAS method (BERTALANFFY and LEBLOND, 1955) (Fig. 28). This property enabled us to demonstrate the outline of the capillary network in "whole membrane preparations" of interalveolar septa regardless of the state of filling of the capillary lumen, and made it unnecessary to fill the capillaries with an injectate. "Whole membrane preparations" of interalveolar septa were easily obtained by sectioning paraffin-embedded blocks of lung tissue at 50 to $150\,\mu$. Such sections always provided unsectioned facets of interalveolar membrane lying normal to the axial beam of light in the microscope, i. e., parallel to the plane of observation. Because of the random arrangement of these facets in space these parallel facets provided randoms amples of the capillary network.

The mode of optical demonstration of the capillary outline is illustrated in Fig. 29. Since the intensely stained basement membrane forms a more or less cylindric tube around the capillary endothelium, light beams passing tangentially (t) through this membrane will have more of their light absorbed than will central beams (c). This results in darker purple outlines of the basement membrane tube of the capillaries. The peak of intensity would theoretically correspond to the internal surface of the tube. The basement membrane, however, is too thin (less than 0.1 μ) (see below) to allow such a refined

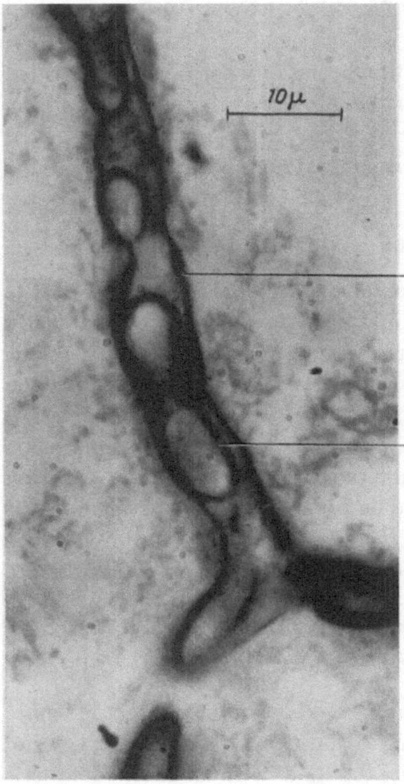

Fig. 28. Demonstration of capillary (C) and epithelial (E) basement membranes by periodic-acid-Schiff reaction

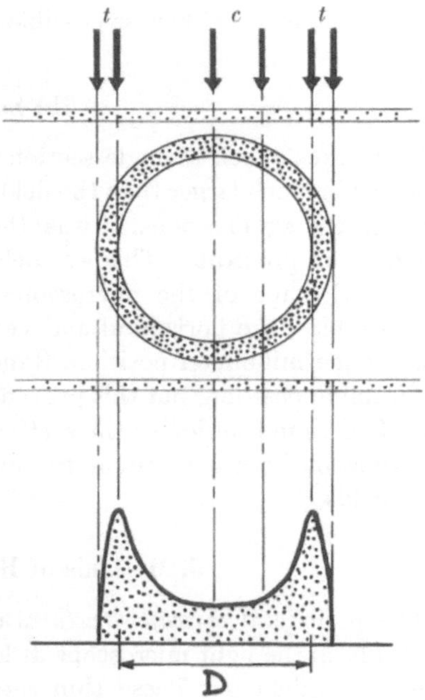

Fig. 29. Light absorption in stained basement membrane permits delineation of capillaries in flat aspect of inter-alveolar septa (compare Figs. 49 and 50)

differentiation. The procedure of measurement and the significance of the dimensions will be discussed on p. 78. The picture of the capillary network obtained by this method is illustrated in Fig. 49.

C. Preparation of Lungs for Electron-microscopic Studies

The high resolution obtained with the electron microscope introduces two special problems into this study. First, optimal preservation of tissue fine structure is required. This cannot be expected from autopsy material. The preliminary studies reported here were, therefore, carried out on fresh animal material. For comparison purposes, human lung tissue was obtained from surgical biopsies; however, the unavoidable manipulation of this tissue caused distortions which do not allow us yet to carry out significant measurements on such material. Second, at the high magnifications of 4000 × to 20,000 × only extremely tiny samples could be analyzed, so that rigorous sampling procedures were essential.

1. Fixation and Processing of Tissue

The animal lungs were fixed under deep anesthesia, achieved by intraperitoneal injection of Nembutal. A tracheotomy was performed; the lungs were collapsed by bilateral intercostal incisions, and a small amount of fixative was immediately instilled without pressure into the trachea. The volume of fixative instilled was chosen in such a fashion that about one third to one half of the lung became filled. In rats about 1.5 ml sufficed. The fixative used here was a 1% solution of OsO_4 in 0.1 molar potassium phosphate buffer at pH 7.4. A few minutes after instillation the chest was opened and the thoracic organs were carefully removed in toto. The OsO_4-fixative had caused a brown discoloration and a slight stiffening of those portions of the lung which were filled. These areas were dissected out with a razor blade, sliced, and carefully cut into small blocks under a drop of fixative. A total of at least 100 to 200 such blocks was obtained, fixed on ice for 2 hours, dehydrated, and infiltrated with Epon by the usual procedure (LUFT, 1961). The sample blocks were then chosen (see below), transferred to capsules, and embedded in Epon.

2. Discussion of Artifacts

The method of fixation used does not ensure a controlled degree of inflation, nor does it avoid distortion of the architecture due to fluid in the airways. It is, however, estimated that such distortions may have little influence on the dimensions that must be obtained.

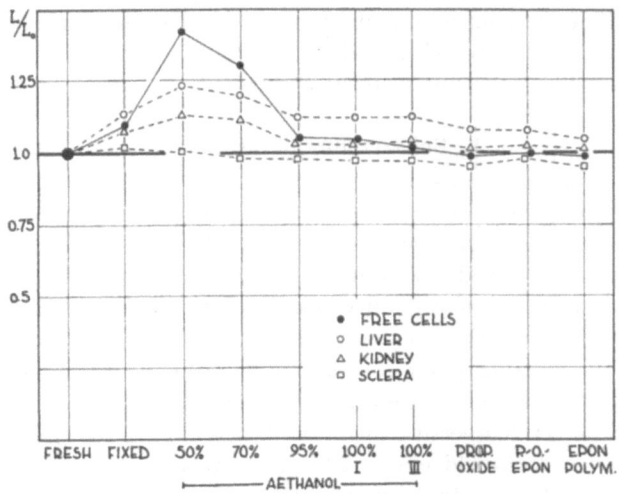

Fig. 30. Experimental determination of change in dimensions of free cells cultured in suspension and different tissues occurring in the course of preparation for electron microscopy

A series of preliminary experiments was carried out to determine dimensional changes that might occur in the course of fixation, dehydration and embedding. The tissues investigated included free cells cultured in suspension, small blocks of liver, kidney and sclera. It was found that cells suffered a considerable swelling in the first stages of dehydration (50% Aethanol); this was less marked in liver and kidney, and was absent in sclera (Fig. 30). With further dehydration and embedding, the original size of cells and tissue blocks was restored, so that the dimensions of embedded structures would presumably compare well with those in the fresh tissue. No correction factors were therefore needed at this stage.

3. Practical Sampling Procedures

In this method, it is likewise necessary to sample in two stages. In the first stage sample sections are to be obtained; in the second stage subsample fields should be chosen from among these sections. Because of the extremely small size of the fields examined, rigorous random sampling rules must be observed.

a) Primary Sampling of Blocks

Before transfer into capsules, the 100 to 200 blocks previously fixed and processed are placed in a petri dish. They are arranged in rows of ten over a piece of square paper whose squares are numbered accordingly. From a table of random numbers n figures are picked; the blocks lying on the corresponding squares are transferred into capsules numbered 1, 2, ..., n. These blocks form the primary sample. Again it is advisable to draw a reserve second sample numbered $n + 1$, $n + 2$, ..., $2n$ as a safety measure.

The blocks of the primary sample are sectioned on a regular ultramicrotome. The block face should be kept as large as possible, not smaller than 1 mm². The sections should not be cut too thin, since the investigation is carried out on the fluorescent screen and at low magnifications; light golden sections (≈ 900 Å) will be satisfactory. The sections are mounted on 400- or 300-mesh grids with a carbon film as support, and stained with lead (KARNOVSKY, 1961). From each primary sample block only one grid is used for measurements.

b) Subsampling of Sections

For this investigation we used a Siemens Elmiskop I. The subsampling of the sections could, therefore, be carried out on the intermediate screen, on which a good portion of the fine mesh grid can be seen with defocussed condenser. Those squares of the fine mesh grid which are accessible to examination are numbered consecutively on a master sheet, an example of which is shown in Fig. 31. A marginal square, such as the one pointed out by an arrow (# 293), is identified and chosen as the starting point for the enumeration of the other squares by moving the grid stepwise across the center hole of the intermediate screen.

With a small objective aperture introduced, it is easy to recognize the location of the section by a dark-field effect. This is sketched on the master sheet, as indicated on Fig. 31. From a table of random numbers n figures are picked which locate the subsample fields; they are marked on the master sheet. It is obvious that fields lying outside the section are disregarded. The subsample fields are now located on the grid in the intermediate screen by enumeration. With brief experience this becomes quite a rapid procedure, since it is not necessary to go back to the point of origin every time. If the subsample fields contain interalveolar septa, the condenser is focussed and the necessary measurements are obtained on the main screen. If they do not contain significant structures, they are discarded and replaced by the next figure on the random number table.

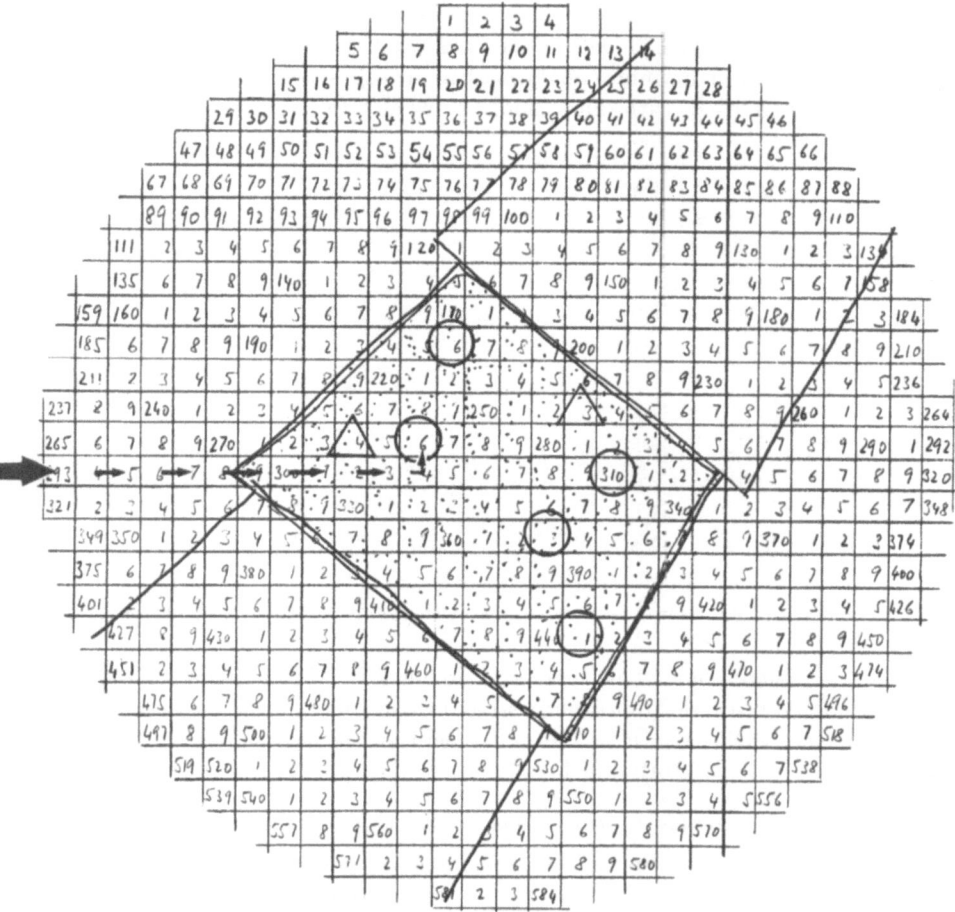

Fig. 31. Master sheet for random number sampling of 400 mesh grid carrying tissue section. Five fields were investigated (circles); two (triangles) were rejected since they did not contain tissue. Fields falling outside section area are rejected a priori. Compare text

Chapter V

Volumetric Composition of the Human Lung

The human lung contains a variety of structures, which were briefly described in Chapter II. They are distributed throughout the organ in a more or less "random" fashion (see p. 10). At various points of the subsequent analysis we shall refer to the relative volumes occupied by these components. We shall, therefore, first attempt a volumetric analysis of the composition of the human lung.

The methods of "point counting" and "linear integration", outlined in paragraph III, B, 1, provide the means by which this analysis can be carried out on two-dimensional samples of lung tissue at any level of magnification. The dimensions of the various components of the human lung range from several millimeters (bronchi, large vessels) to the order of 60 Å (cytoplasmic membranes of epithelial and endothelial cells). While the larger structures can be observed with the naked eye, investigation of the smaller components which constitute the respiratory portion of the lung (alveoli, alveolar ducts and capillaries) requires the use of an optical microscope. Furthermore, analysis of the composition of the interalveolar septa at the cellular level, with dimensions as low as 60 Å, will necessitate the use

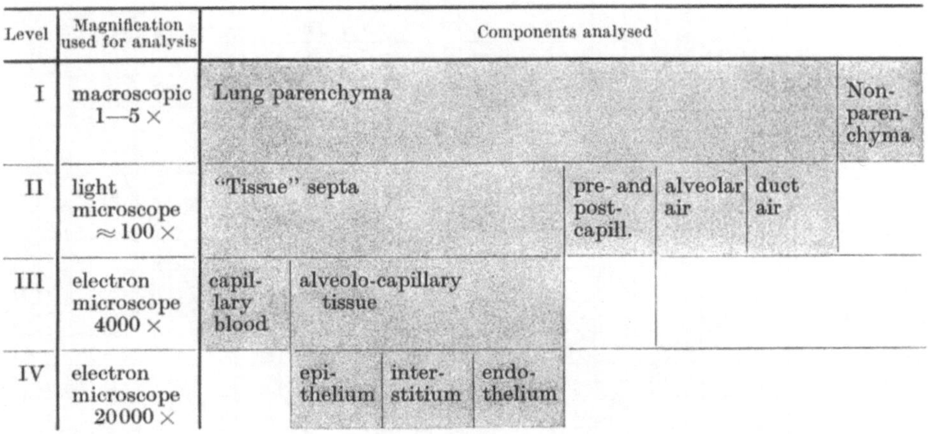

Level	Magnification used for analysis	Components analysed						
I	macroscopic 1—5 ×	Lung parenchyma						Non-paren-chyma
II	light microscope ≈ 100 ×	"Tissue" septa			pre- and post-capill.	alveolar air	duct air	
III	electron microscope 4000 ×	capil-lary blood	alveolo-capillary tissue					
IV	electron microscope 20000 ×		epi-thelium	inter-stitium	endo-thelium			

Fig. 32. Graphic representation of the components analyzed volumetrically at different levels of magnification

of an electron microscope. It is, therefore, obvious that the volumetric analysis of an lung should be carried out in successive steps at different levels of magnification, each one chosen because it is more convenient for the particular investigation. This is schematically illustrated in Fig. 32.

Level I: Fractional volumes of parenchyma[1] and non-parenchymatous structures

Non-parenchymatous structures of the lung are defined as conductive airways, conductive blood vessels, and coarser connective tissue structures, e. g., interlobular septa, peribronchial and perivascular spaces, and pleura. On cut surfaces of the fixed lung these structures can easily be distinguished from the "spongy" respiratory portion of the lung which we will call "parenchyma"[1].

The fixed lungs were sectioned into parallel slices of approximately 1.5 cm thickness, as described in Chapter IV. Photographic records of the cut surface of all slices of two of the lungs (age 34 and age 48) were used for evaluating the fractions φ of parenchyma and $(1 - \varphi)$ of non-parenchymatous structures by point counting. A hexagonal lattice of points was superimposed on large prints, as illustrated in Fig. 33, and a differential count of the points lying on parenchyma or non-parenchymatous structures was made.

---

[1] In this context the term "parenchyma" refers to that portion of the lung which contains alveoli and capillaries. As a purely descriptive term it means the "spongy" tissue observed on cut surfaces of fixed lungs.

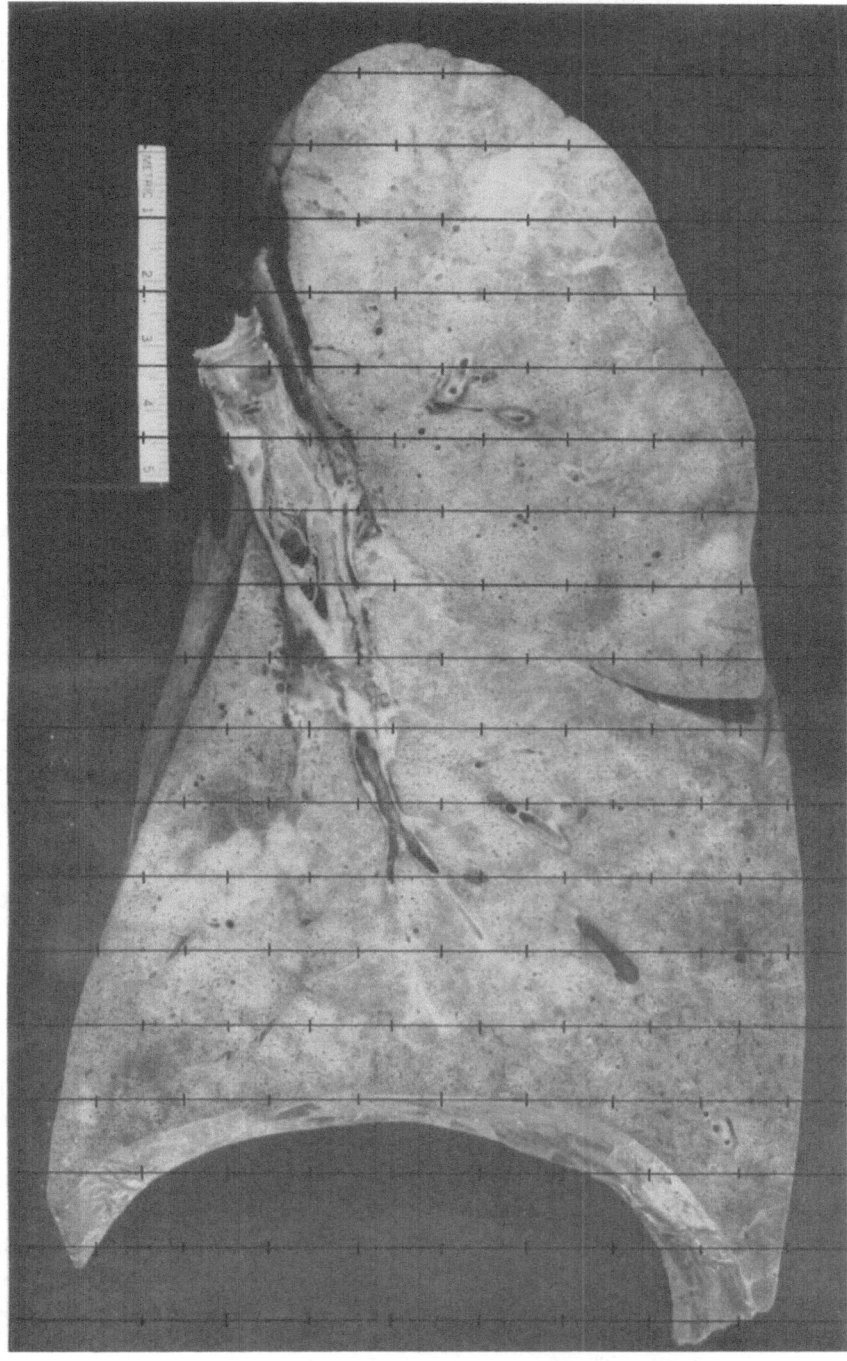

Fig. 33. Cut surface of one of the slices of the lung of a young woman. A hexagonal lattice of points has been superimposed at random, thus allowing volumetric analysis by point counting. Compare text
[From WEIBEL, 1962]

The findings in these two cases are tabulated in Table V.1. The parenchyma was found to occupy about 90%, the non-parenchymatous structures about 10%, of the total lung volume. The standard deviation of φ refers to the variation of the differential counts obtained on each slice; it was roughly 4%. The 95% confidence interval is of the order of ±1.5% in both cases.

Table V.1. *Volumetric Proportion between Lung Parenchyma and Non-Parenchymatous Structures*

#	Parenchyma φ S. D.	Non-parenchyma $1 - \varphi$	95% Confidence Interval for φ	Number of points
9975	.901 ± .040	.099	±0.013	2300
8710	.909 ± .036	.091	±0.015	1600
Average	.9	.1		

Level II: Fractional composition of parenchyma

The "respiratory portion" or parenchyma is defined as that part of the lung which contains the structures of the respiratory and transitory zone described on p. 9, i. e., alveoli, alveolar capillaries, and alveolar ducts and sacs. It also contains some delicate tissue which is associated with the capillaries in forming the interalveolar septa. These structures can be conveniently resolved at the medium magnifications provided by an optical microscope. Their volumetric distribution was, therefore, analyzed at a magnification of $100 \times$ by the linear integration method of SCHUCHARDT (1954), as discussed on p. 17. The results of this analysis are listed in Table V.2 and represented graphically in Fig. 34 with respect to the age of the individuals. The volumetric fraction "alveolar air" (ϱ_A) comprised on the average 62% of the parenchyma in the three younger individuals. This fractional volume appeared to be reduced to 56% in the two older persons. Con-

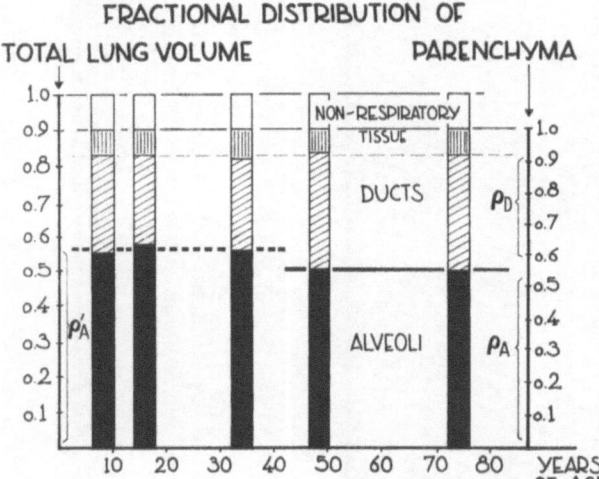

Fig. 34. Fractional distribution of the total lung volume and of the volume of pulmonary "parenchyma" among alveoli, alveolar ducts, and "tissue". A shift in the distribution of the air space between alveoli and ducts occurs with age [From WEIBEL, 1962]

versely, the fractional volume of the central air ducts (ϱ_D) was 29% in the younger subjects and increased to 37% in the older individuals. The small number of cases obviously does not justify conclusive statements with respect to changes of volume proportions of the lung generally attributable to age. For these five cases, however, the difference between the two age groups was found to be statistically significant ($P < 0.001$). The fractional volumes of parenchymal tissue (τ) and transitory blood vessels (v) averaged 7% and 1.3% respectively; no significant changes associated with age could be detected. As discussed on p. 24, the thickness of the histologic sections used in this analysis had caused a significant over-estimation of the tissue component τ, with a corresponding underestimation of the fraction ϱ_A. The values of τ and ϱ_A presented in Table V.2, and Fig. 34 have been corrected according to equations (3.35) and (3.36), respectively.

Table V.2. *Volumetric Proportion of Components of Parenchyma*

#	Alveoli* ϱ_A	S. D.	Duets ϱ_D	S. D.	"Tissue"* τ	Vessels v
2471	.61	± .042	.31	± .057	.066	.014
555	.64	± .034	.28	± .026	.073	.010
9975	.62	± .041	.28	± .041	.079	.014
8710	.56	± .051	.37	± .081	.060	.012
3361	.55	± .055	.37	± .087	.066	.012

* After correction for influence of section thickness [equations (3.35) and (3.36)].

Levels III and IV: The component "tissue" of level II is composed of alveolar capillaries, endothelial and epithelial cells, and interstitium (Fig. 35). So far, only preliminary estimates of these fractions can be offered. Capillary blood is found to

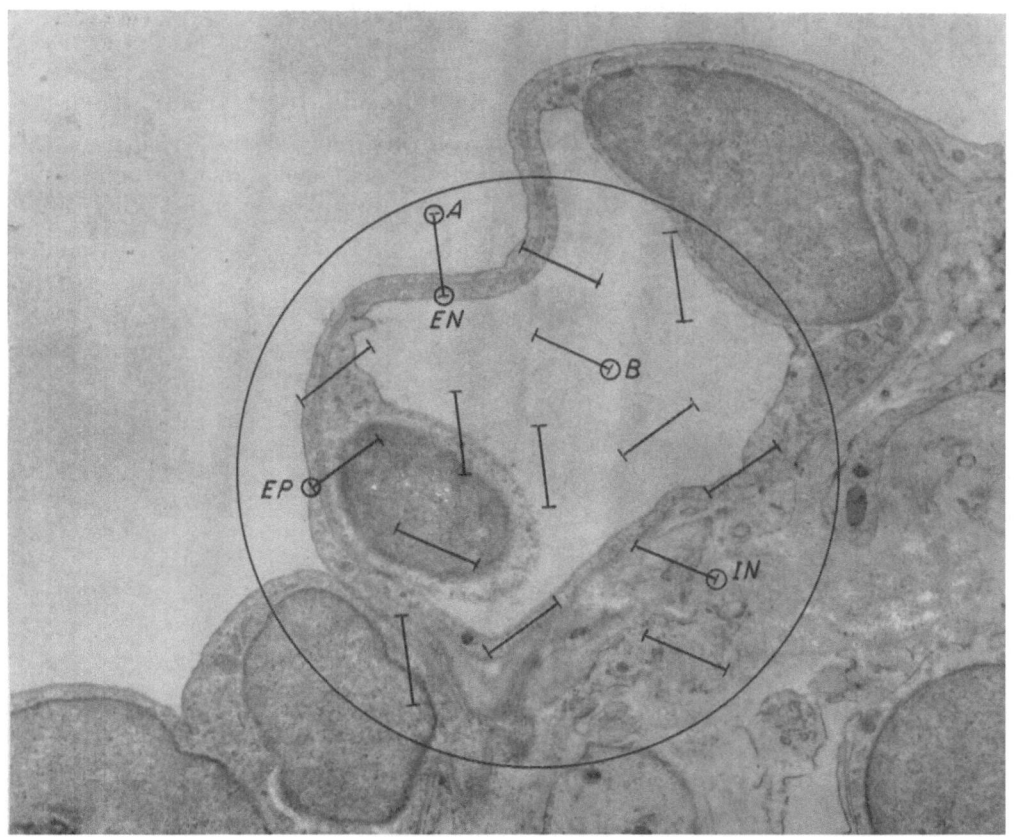

Fig. 35. Electron micrograph of alveolar capillary of rat lung with template of lines of multipurpose measuring screen (cf. Fig. 21) superimposed at random. The end points of the 15 lines lie on the vertices of the equilateral triangles of a hexagonal lattice and can thus be used for volumetry at levels III and IV. Five points which lie on the discernible components are designated by circles: *A* alveolar air, *B* capillary blood, *EP*: epithelium, *EN*: endothelium, *IN*: interstitium. At the bottom a small epithelial cell is shown . × 10,000

amount to about 50% of τ, which corresponds to 3% of the entire lung volume. The remaining 50% consists of tissue in the actual sense, where of 30% comprises endothelial and epithelial cells, while 20% is formed by the interstitium. The distribution of the components of the actual tissue space will be discussed further on p. 103.

Chapter VI

Geometry and Dimensions of Airways of the Respiratory Zone

The pulmonary "parenchyma", as it was defined above, contains the elements of the respiratory and transitory zones. In the preceding chapter it was shown that the alveoli were by far its largest component. From the viewpoint of respiratory function, alveoli are of primary interest since they represent the air chambers in which the gas exchange between air and blood actually takes place. The following paragraphs will, therefore, deal with the study of geometry, number and dimensions of alveoli, and with the determination of some over-all dimensions, such as the total alveolar surface area, which is an estimator of the size of the air-tissue-interface.

A. Geometry of Alveoli and their Relationship to Alveolar Ducts

MALPIGHI (1697) first recognized the alveoli (vesiculae) as the terminations of the airway system which were separated only by thin membranous structures containing the "rete mirabile" of the capillary network. He described them as

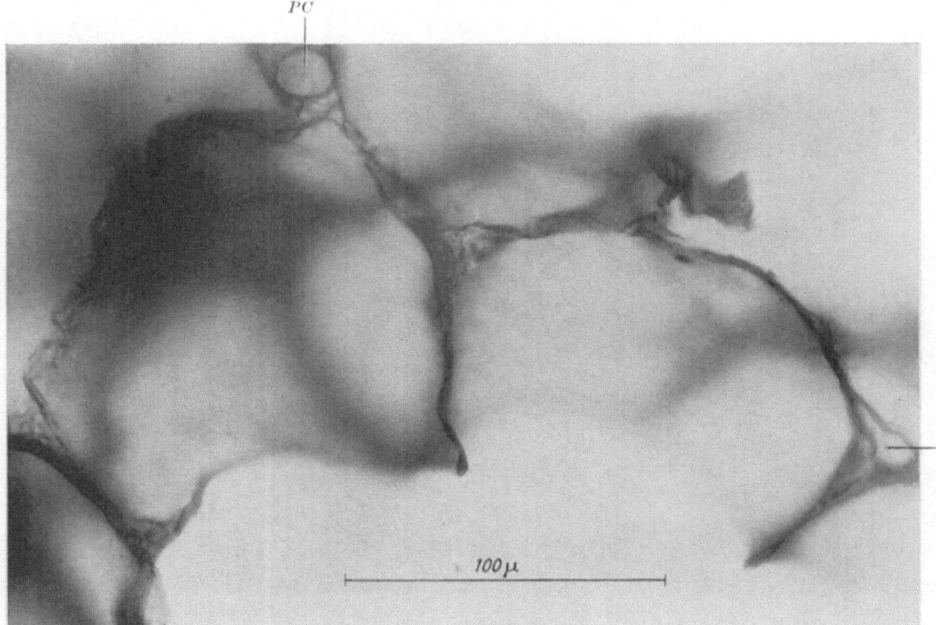

Fig. 36. Thick section of human lung showing two alveoli which open on alveolar duct. Note the cup-like shape of the alveoli, and compare with the model of Fig. 37. PC = precapillary. × 330
[From WEIBEL, 1962]

comparable to the cells of a honeycomb (see below), but stated that they communicated with each other in a sponge-like fashion. It was not until the quite recent work of SCHULZE (1871) and LAGUESSE (1901) that the proper relationship

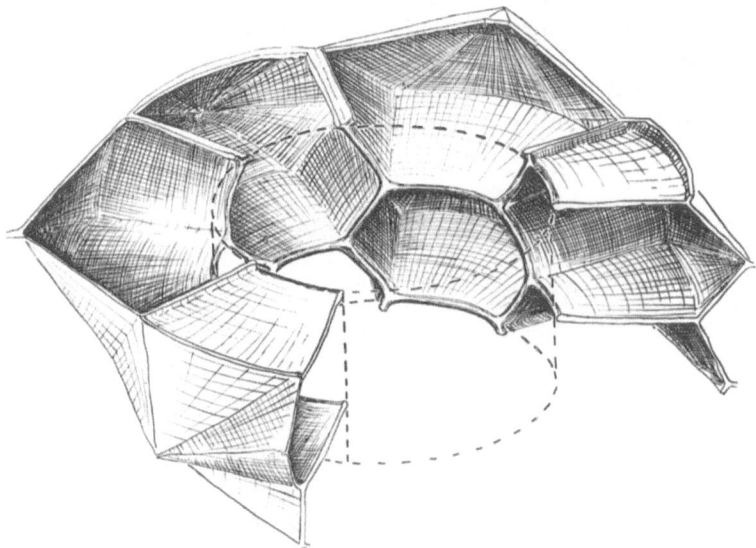

Fig. 37. Schematic model of arrangement of alveoli around alveolar duct. Compare with photomicrograph of Fig. 36 and with honeycomb illustrated in Fig. 38 [From WEIBEL, 1962]

Fig. 38. Photograph of cells of honeycomb. Cells have hexagonal cross section. The three crests of the domes correspond to the lateral walls of the cells of the opposite layer

of alveoli to the elements of the transitory zone (ductuli et bronchioli alveolares) was established. This relationship, however, together with the close packing of alveoli against one another, determines the geometry of these peripheral air chambers.

Alveoli are small lateral outpouchings of alveolar ducts, alveolar sacs and repiratory bronchioles. MACKLIN (1943) described them as cup-like and represented

them in a model which quite appropriately resembles the picture of a single
alveolus as it can be observed in thick sections of inflated lungs (Fig. 36). However, this model has a smooth wall and does not indicate the relationship of this
alveolus to its neighbors. The analogy to the cells of a honeycomb has therefore
often been suggested (MALPIGHI, 1697; ELZE and HENNIG, 1956; and others),
taking into account the close packing of alveoli. This relationship is schematically
represented in Fig. 37. Upon comparison with the photograph of Fig. 38 it is
evident that a honeycomb can serve as an acceptable alveolar model if one

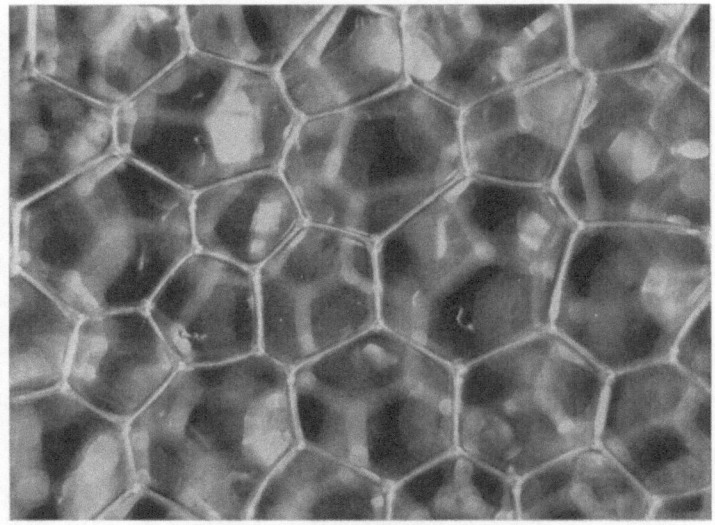

Fig. 39. Photograph of bubbles of a soap foam. Note that three facets of the film always meet despite irregularity
in size and shape of bubbles

imagines it bent around the cylindric surface of the alveolar duct. The hexagonal
prism of the honeycomb cell is then transformed into a pyramidal wedge of the
air sleeve surrounding the duct.

The dome of the honeycomb cell consists of three equilateral rhombi which
meet at the characteristic Maraldi angle of 109° 27′ (cos⁻¹ 1/3) (D'ARCY THOMPSON,
1942). This is due to the packing of two layers of cells in which the vertices of the
cells of one side coincide with the line of junction of three lateral facets of the
other side (Fig. 38). The dome of an alveolus, by contrast, shows a more complex
structure. It is built mostly of more than three facets; this results from close
packing, with several alveoli belonging to adjacent alveolar ducts. Accordingly,
each alveolus may have more than one vertex. Furthermore, size and shape of the
facets vary considerably. The picture, thus obtained, more nearly resembles that
of closely packed bubbles of foam, which have an irregular polyhedral configuration (Fig. 39).

The partitions of honeycomb and foam can be subdivided more or less into
plane facets which separate two adjacent chambers. In general, three such facets
join in a common line of junction at an average angle of 120°. This is also true for

alveoli. Fig. 40 shows a thin section of five interalveolar facets which meet in two lines of junction, and separate four alveoli which are partially seen in this photograph. This arrangement is also apparent in Fig. 36, although it is partially obscured in this thick section due to the inclination of some of the facets to the plane of section. In the fresh lung, these facets can be assumed to be more or less flat if there are no pressure differences between neighboring alveoli. The slight distortion of their image on histologic sections is an artifact.

From the preceding discussion it follows that alveoli can vary considerably in configuration and that any definition of their geometry should be kept in

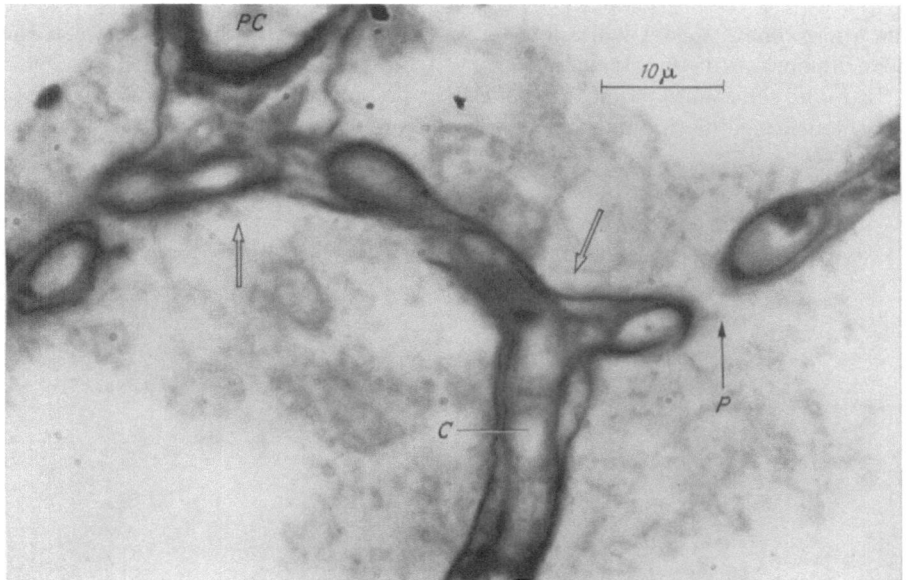

Fig. 40. Two junctures of three interalveolar septa (arrows) in human lung. The septa which are covered by epithelium contain alveolar capillaries (C) and precapillaries (PC) and are occasionally perforated by interalveolar pores (P) through which the epithelium of adjacent alveoli is continuous

general terms. In the following presentation we shall, therefore, use a few simple geometric models which may represent one or the other characteristic of a"typical" alveolus. We shall then seek a significant parameter which is relatively insensitive to variations in configuration and should be able to represent the surface-to-volume ratio of the alveolus.

Alveoli form a sleeve around the alveolar duct; they can be regarded as simple wedges of a hollow cylinder (Fig. 41 a and b). These wedges have either a hexagonal or an elliptic cross section. In first approximation their dome corresponds to the cylindric surface outlined by the lateral wall of the wedges; in better approximation this dome is replaced by an elliptic cone (Fig. 41 c). This model alveolus comes quite close to reality. In order to facilitate its geometric description, the elliptic cones of body and dome are replaced by simpler circular cones. If we fix the opening angle of the truncated cone of the "body" as being $\tan^{-1} 1/2$ such a model alveolus is defined by three dimensions (Fig. 41 d): the "mean" radius r of the body, the depth of the dome a, and the total depth A. Surface s'_A and volume v'_A of this model are given by:

$$s'_A = \pi \left[2\sqrt{1 + 1/4}\ (A - a) \cdot r + q \cdot a \sqrt{1 + \left(\frac{q}{A}\right)^2}\ \right] \qquad (6.1)$$

$$v'_A = \pi \left[p^2(A - a) + \frac{1}{2}\, p(A - a)^2 + \frac{1}{12}\, (A - a)^3 + \frac{1}{3}\, q^2 a \right] \qquad (6.2)$$

where

$$p = r - \frac{A - a}{4} \qquad (6.3)$$

$$q = r + \frac{A - a}{4} \qquad (6.4)$$

This approximate model will allow us to estimate the relationship between the linear dimensions of the alveoli.

It may, sometimes, appear useful to represent the alveolus by a body of smooth surface. For this purpose it is best to replace the wedge-shaped alveolus

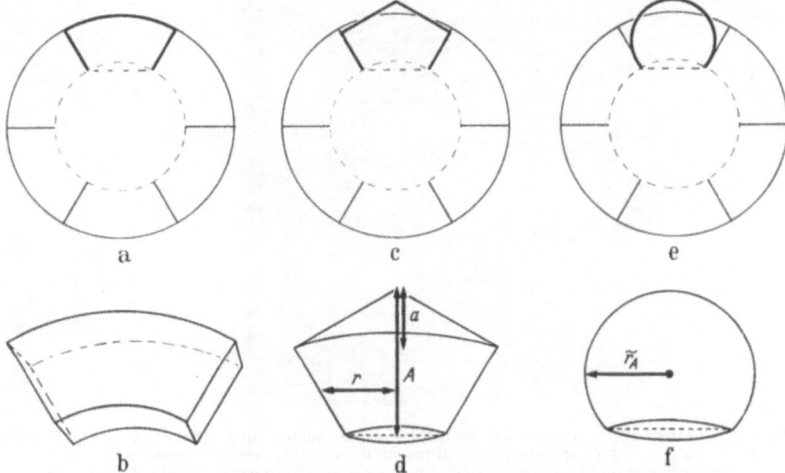

Fig. 41. Different geometric simplifications of the alveolar shape used in this study. Compare text

by 5/6 of a sphere, as shown in Fig. 41 e and f. This simplest model is quantitatively characterized by a single dimension, the average alveolar diameter d_A.

The only geometric property of the alveolus which will enter the subsequent calculations is its surface-to-volume relationship. It is conveniently expressed by the dimensionless coefficient

$$\bar{\sigma}_A = \frac{s_A}{v_A^{2/3}} \qquad (6.5)$$

which was introduced on p. 36. Since no well-defined geometric form has been found to represent the "typical" alveolus, the determination of $\bar{\sigma}_A$ remained, at first, an empiric estimate. It was necessary to consider that the alveolus was open on one side and had the somewhat wedge-shaped polyhedral configuration described above. The coefficient $\bar{\sigma}_A$ was, therefore, computed for various bodies which would have some of the properties of the alveolus, such as open cube,

5/6-sphere, wedge of a hollow cylinder, dodecahedron with one side open, and the double-cone model already described (Fig. 41 d). In addition, cardboard models of "typical" alveoli were constructed from stereomicroscopic examinations and measured. From these different approaches it was empirically estimated that $\bar{\sigma}_A \approx 4.8$ would most appropriately represent the surface-to-volume relationship of the alveolus.

Equation (3.94) on p. 39 finally provided the means for calculating the actual value of $\bar{\sigma}_A$ (Table VI.1). In our five lungs it was found to be

$$\bar{\sigma}_A = 4.87 \text{ [S. D. } \pm 0.19 \text{ (4\%)]}.$$

By introducing the empirically estimated value of $\bar{\sigma}_A = 4.8$ into equation (3.91), we have introduced an insignificant error of about -1.5%.

Table VI.1. *Surface-to-Volume Ratio σ_A of Alveoli from equation (3.94)*

#	σ_A
2471	4.84
555	5.04
9975	5.02
8710	4.56
3361	4.92

$\bar{\sigma}_A = 4.87$ [S. D. ± 0.19]

In addition, we needed an estimate of the coefficient β_A relating the mean cross-sectional area of the alveolus to its volume [equation (3.39)]. This coefficient, however, is quite insensitive to possible variations in configuration (Fig. 17). Regarding the alveolus as a polyhedron circumscribed to a sphere, the coefficient β_A was empirically estimated to have a value of about 1.55 (WEIBEL and GOMEZ, 1962).

B. Number of Alveoli

1. Results of our Studies

The number of alveoli could be estimated in our five normal human lungs by applying the principle of direct counting outlined on p. 25 (WEIBEL and GOMEZ, 1962a). Equation (3.40) yields the number of alveoli in the unit volume of lung parenchyma. Its calculation requires knowledge of the coefficient of configuration β_A, of the volumetric density of alveoli ϱ_A, and of the number n_A of alveolar transsections on the unit area of a thin histologic section. The coefficient β_A was estimated in the preceding paragraph to be 1.55. The volumetric fraction ϱ_A is listed in Table V.2, for all five cases. The number of transsections per unit area was counted on random histologic sections by following the procedure outlined on p. 28. From the counts obtained on fields of $9.1 \cdot 10^{-3}$ cm² the number per 1 cm² was calculated (Table VI.2). It was found to be inversely proportional to the size of the lungs. The 95% confidence interval of these counts averaged 5% of the

Table VI.2. *Number of Alveoli*

#	n_A/cm^2 .10³	S. D.	95% Confidence interval (% of mean)	N_A/cm^3 .10³	S. D.	N_{AT}
2471	52.4	± 6.45	± 3.5%	322	± 60	304 · 10⁶
555	49.8	± 7.10	± 4%	292	± 72	288 · 10⁶
9975	31.1	± 7.24	± 6%	148	± 54	310 · 10⁶
8710	24.8	± 4.97	± 6%	110	± 31	286 · 10⁶
3361	27.9	± 4.56	± 5%	131	± 31	286 · 10⁶

n_A = number of alveolar transsections (average),
N_A = number of alveoli, (average),
N_{AT} = total number of alveoli.

mean count, so that our confidence requirements are well satisfied (cf. p. 14). By means of equation (3.40) the average number N_A of alveoli per 1 cm³ of lung parenchyma was calculated, and the total number N_{AT} of alveoli in the entire lung was obtained from

$$N_{AT} = N_A \cdot \varphi V_{LS} \qquad (6.6)$$

where φV_{LS} is the volume of the lung parenchyma in the processed lung. The numbers N_{AT} are listed in the last column of Table VI.2. It was found that all

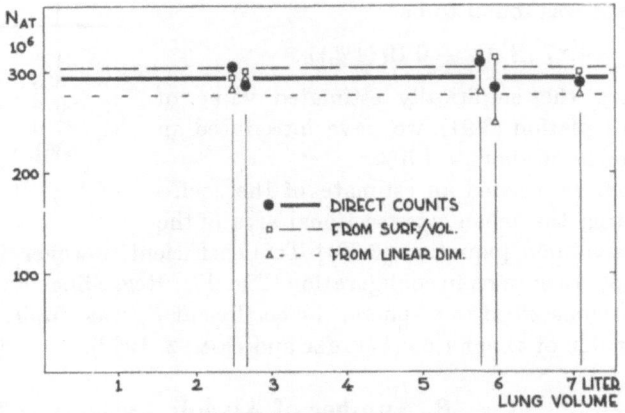

Fig. 42. Total number of alveoli in five normal human lungs as determined by three independent methods plotted as function of lung size

five lungs contained essentially the same number of alveoli despite the great variation in size and age of these lungs. The group average was 296 millions, with a standard deviation in the individual cases of only 4%. The 95% confidence interval of this group mean is about ± 13 millions. In Fig. 42 the values of N_{AT} are plotted as a function of lung size; no dependence is apparent.

The total number of alveoli could, in addition, be inferred indirectly by dividing the total alveolar volume $\varrho_A \cdot \varphi V_L$ by the average individual volume $\bar{v}_A$ of an alveolus. The average volume $\bar{v}_A$ could first be estimated in an approximate manner by introducing the linear dimensions A, a and r discussed below into equation (6.2). The values of N'_{AT} thus obtained are indicated in Fig. 42 as small triangles; these rough estimates are, on the average, somewhat lower than the figures obtained by direct counting, but the group average of 270 millions is still of the same order of magnitude. A third way of estimating $\bar{v}_A$ was proposed by ELZE and HENNIG (1956), who found that the surface-to-volume ratio of the alveolus was of the same order of magnitude as that of a *closed* sphere. Having determined the alveolar surface area S_{AT} by the method of "mean chord length" (see p. 37), they calculated the diameter d_{SA} of the spheric analog to the alveolus from

$$d_{SA} = 6 \cdot \frac{V_{AT}}{S_{AT}} \qquad (6.7)$$

where $V_{AT} = \varrho_A \cdot \varphi V_L$ is the total alveolar volume. The volume $\bar{v}_A$ of this spheric analog can be computed from d_{SA}, and the number N'_{AT} is thus derived. But this procedure can be simplified if the surface-to-volume ratio of the spheric analog is again expressed by the dimensionless coefficient σ introduced on p. 36. The value of σ_s for a sphere of any size is

$$\sigma_s = \sqrt[3]{36 \cdot \pi} = 4.834 \ldots .$$

(Note that σ_s is quite similar to $\bar{\sigma}_A$ given on p. 61). We, thus, obtain directly

$$N''_{AT} = \frac{S^3_{AT}}{36 \pi \cdot V^2_{AT}} \, . \tag{6.8}$$

The values of N''_{AT} calculated by (6.8) for our five lungs are plotted as open squares in Fig. 42; again, N''_{AT} apparently does not depend on the lung size. The group average of 304 million falls within the 95% confidence interval of the direct counts.

2. Discussion of Results

In conclusion, we ascertained by three methods that the total number of alveoli in these five normal human lungs was independent of the lung size, with an average of approximately 300 million. The youngest of these lungs was 8 years old; it appears, therefore, that the adult total number of alveoli is reached before that age. In subsequent development the lung grows through increasing alveolar dimensions (see below).

The estimate of 300 million alveoli appears to conform reasonably with widely quoted figures (POLICARD, 1938; MÜLLER, 1945; VON HAYEK, 1953) which revert to the older studies listed in Table I. However, the figure differs significantly from the result of more recent studies by ELZE and HENNIG (1956) and HIERONY-MI (1961) who estimated the total number of alveoli to be approximately 70 millions. While earlier investigators had sometimes used questionable methods, these more recent results were obtained from determinations of the "mean chord length" (TOMKEIEFF, 1945; HENNIG, 1956), as described on p. 37. This quite reliable method has also been used in the present study. The sizable difference in results must, therefore, be due to other errors. ELZE and HENNIG used data obtained by KULENKAMPFF (1957) on one lung, which was fixed in situ in a very low degree of inflation by instillation of formalin into the trachea. This procedure is likely to cause uneven distention of the lung since it does not allow the reopening of the many atelectases present in a cadaver lung. The instillation of a fluid may exaggerate this unevenness by trapping the air in those acini which were not atelectatic and by pushing it further toward the periphery. These parts of the lung will then be distended while atelectatic regions remain collapsed. Considering that about 90% of the parenchyma of an inflated lung is air, it is obvious that inflated alveoli can occupy a volume up to 9 times as large as an equivelent number of collapsed alveoli. This means that the chance of including distended alveoli in the sample is much greater, so that the total alveolar number obtained will fail to include alveoli in collapsed areas. This error may be exaggerated if "suitable" areas of the lung are sampled. In addition, the degree of preparative shrinkage from fixation to slide (factor p in Table IV.2) was estimated by KULEN-KAMPFF to be 40%, which appears to be much too large compared with the factor $p = 9\%$, determined experimentally in this study for paraffin embedding. Considering these two essential points, it is quite possible that ELZE and HENNIG may have underestimated the number of alveoli. HIERONYMII appears to have studied several lungs by adopting the same procedure, criteria and constants that HENNIG used, so that the same criticism applies. Since he does not present statistics on his results, the latter are difficult to evaluate. In conclusion, we believe that these two studies cannot invalidate our results, since, first, two important sources of error were carefully avoided by us, and, second, the results of different procedures on five different lungs yielded consistent results which satisfied specified confidence requirements.

C. Linear Dimensions of the Average Alveolus

On p. 60 we defined a simplified model alveolus (Fig. 41d) whose "body" consists of the shell of an inverted truncated cone of average radius r; it is closed

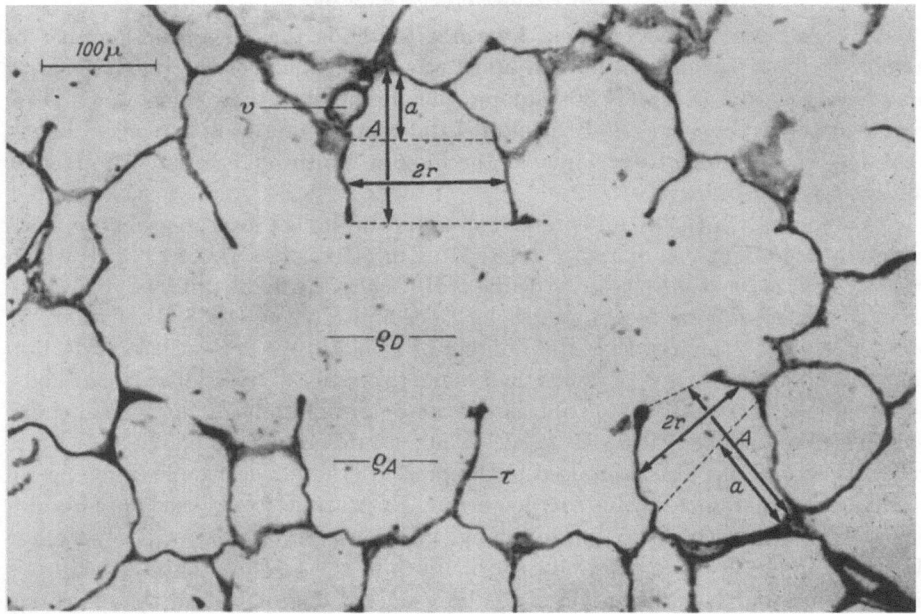

Fig. 43. Oblique section of alveolar duct of human lung with sections of surrounding alveoli. The dimensions A, a, and $2r$ are sketched into two alveoli as examples. Compare Fig. 41d

at its larger cross section by a cone-shaped dome of depth a. The total depth A of the alveolus is measured from the entrance plane to the deepest point of the dome. These dimensions were estimated on thin histologic sections by direct microscopic measurement with an eyepiece micrometer, as described on p. 29. Clear-cut cross sections or longitudinal sections through alveolar *ducts* were selected in order to make possible identification of the three well-defined dimensions A, $2r$ and a on longitudinal sections of alveoli as illustrated in Fig. 43. In each lung 40 such alveolar sections have been selected and measured, thus providing 120 measurements. The relative frequency distributions of the dimensions A and $2r$, obtained in this fashion on our five lungs, are plotted on Fig. 44. The average dimension calculated from these measurements is indicated as a vertical line. It shifts toward larger values when the lungs are larger.

By obtaining these dimensions on thin histologic sections we were actually measuring dimensions of more or less randomly placed sections through alveoli. The planes of these sections will, therefore, be more or less displaced from the equatorial plane and will not represent the largest cross section of the alveolus. The frequency distributions in Fig. 44 are, therefore, the result of a) a variation

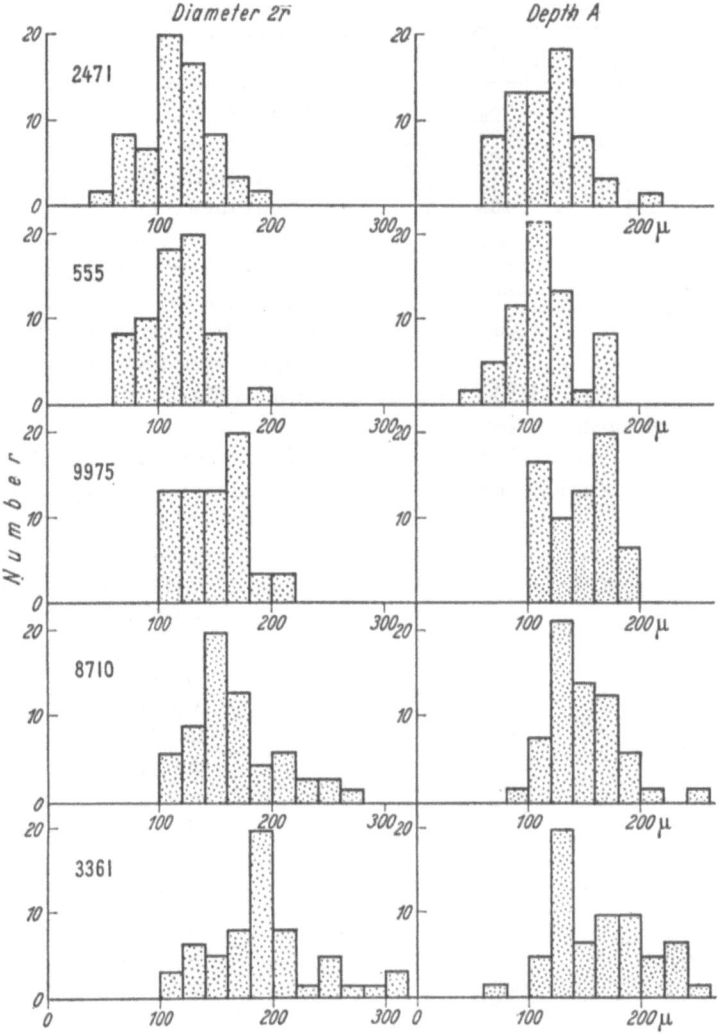

Fig. 44. Distribution of diameter and depth of alveolar sections in the five investigated cases. Compare text for significance of these dimensions

in the true diameter or depth of the alveoli, and b) a varying displacement of the section plane from the "equator" of the alveolus. Thus, the average dimensions obtained from these measurements are too small when compared with the true dimensions, as was discussed on p. 29. WICKSELL (1923) and LENZ (1956) attempted to separate this combination of two superimposed distributions occurring by random sectioning of a set of spheres of unequal size. Using the practical method proposed by LENZ, we estimated that the average dimensions A and r should be multiplied by a factor of about 1.2 to yield the "true" average dimensions of the alveoli in the processed state. The dimensions thus obtained were converted into the fresh state by the corresponding correction factors $1/f_p$ (Table IV.2). It followed also, that the size distribution of the younger cases was rather narrow, while it became wider in the older lungs.

The dimensions $\bar{A}$, $\bar{r}$ and $\bar{a}$ are listed in Table VI.3. Each of the three dimensions increases with the total lung volume. In adults the diameter ($2r$) and the total depth (A) of the average alveolus were of the order of 250 to 300 μ. As a matter of historical interest, it should be pointed out that the Rev. STEPHEN HALES in 1731 had calculated that the average alveolus of a calf's lung measured "1/100 part of an inch" (254 μ), which turns out to be a surprisingly accurate estimate.

Table VI.3. *Linear Dimensions of Alveoli*

#	V_L cm³	$\bar{A}$ 10^{-4} cm	$\bar{a}$ 10^{-4} cm	$\bar{r}$ 10^{-4}	d_{SA} 10^{-4} cm
2471	2580	199	65	99	215
555	2680	193	57	99	223
9975	5750	260	70	128	275
8710	7100	258	73	145	314
3361	5950	281	92	138	275

V_L = total lung volume in fresh state,
$\bar{A}$ = depth of alveolus (see Fig. 10),
$\bar{a}$ = depth of cone-shaped dome,
$\bar{r}$ = mean radius of alveolus,
d_{SA} = diameter of spheric analog.

For purposes of comparison we can define a characteristic linear dimension of the average alveolus as its approximate average diameter $\tilde{d}_A$. This would correspond to the diameter d_{SA} of the spheric analog of the alveolus, which was introduced in equation (6.7) on p. 62, following Hennig's suggestion (Table VI.3). Another estimate of $\tilde{d}_A$ can be obtained by averaging the values of A and $2r$, determined on sections. The values thus obtained have been plotted on Fig. 45 against the lung volume. The heavy dot represents the average of the two estimates. It appears that $\tilde{d}_A$ increases with the lung volume V_L. Since the number of alveoli was found constant for all lungs, as well as the surface-to-volume coefficient $\bar{\sigma}_A$, it is expected that this dependence is of the form

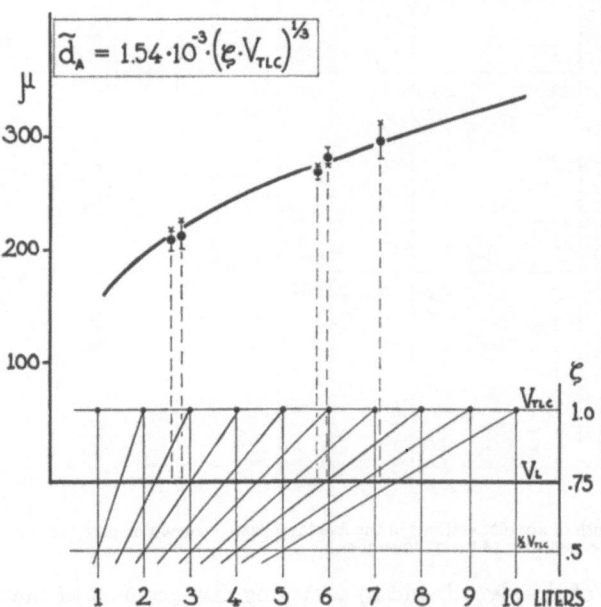

Fig. 45. Average alveolar "diameter" of fresh human lungs as function of lung volume. Isopleths around abscissa represent attempt to relate $\tilde{d}_A$ to different degrees of inflation ζ on hypothetic grounds. Compare text

$$\tilde{d}_A = k \cdot V_L^{1/3} . \qquad (6.9)$$

A satisfactory "fit" of this function to the points is obtained if $k = 1.54 \cdot 10^{-3}$.

The dimensions $\bar{A}$, $\bar{r}$ and $\bar{a}$, and d_{SA}, as well as $\mathring{d}_A$, so far discussed, refer to a lung at about 3/4 of its maximal inflation. The possibility of introducing varying degrees of inflation into this relation will be discussed on p. 72.

D. Total Alveolar Surface Area

The total alveolar surface area S_{AT} represents a first estimate of the over-all area available for gas exchange in the lung. The dimensions of more appropriate exchange interfaces, to be defined later, will be proportional to S_{AT}. It is, therefore,

Methods for alveolar surface

Geometrical Honey comb model	Mean chord (Tomkeieff-Hennig)	Linear dimensions Model of average alveolus

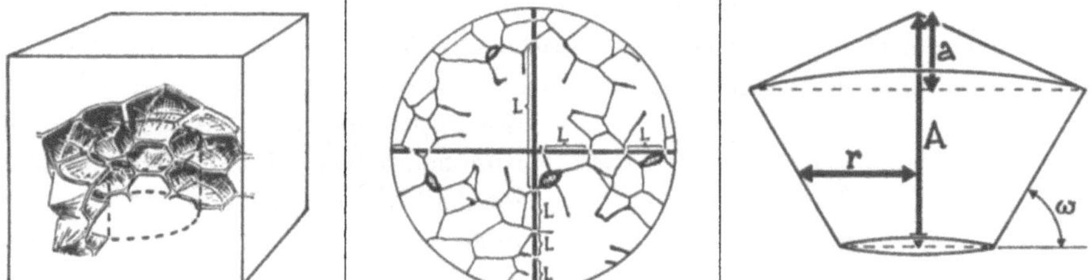

$$S_{AT} = N_{AT} \cdot \bar{\sigma}_A \left(\frac{\varrho'_A \cdot V_L}{N_{AT}} \right)^{2/3}$$

$$S_{AT} = \frac{4 \cdot \varphi \cdot V_L}{\bar{L}_m}$$

$$s_A, v_A : f(r, A, a, w)$$

$$S_{AT} = \frac{\varrho'_A \cdot V_L}{\bar{V}_A} \cdot \bar{s}_A$$

Fig. 46. Three independent methods for determining the dimensions of the alveolar surface area S_{AT}

evident that this essential dimension will need to be assessed with particular care. We have, accordingly, employed three different methods, discussed in detail in Chapter III, in order to arrive at three independent estimates of S_{AT} (Fig. 46).

1. Estimation of Total Alveolar Surface by Geometric Method

On p. 61 we discussed the geometric properties of the normal alveoli with respect to their surface-to-volume relationship. This enables us to use equation (3.93) to estimate the total alveolar surface area: $S_{AT1} = N_{AT}^{1/3} \cdot \bar{\sigma}_A (\varrho'_A V_L)^{2/3}$. The total number of alveoli N_{AT} had been obtained by direct counting (TableVI.2), and the total lung volume (V_L) had been measured by water displacement after fixation (Table IV.1). V_L is introduced in terms of its value for the fresh lung since we are mainly interested in this more or less "physiologic" state. Some later considerations on the capillary network, however, will require the knowledge of the alveolar surface of the processed lung (S'_{AT1}), which is easily determined by introducing V_{LS} into equation (3.93). The alveolar

Table VI.4. *Alveolar Surface by Geometrical Method 1*

#	V_{LS} cm³	S'_{AT} 10⁴ cm²	V_L cm³	S_{AT} 10⁴ cm²
2471	1050	23.8	2580	43.3
555	1090	25.1	2680	45.7
9975	2340	42.6	5750	77.5
8710	2890	44.0	7100	80.1
3361	2420	39.0	5950	71.0

V_{LS}, S'_{AT} = lung volume and total alveolar surface in "processed" lung.

V_L, S_{AT} = lung volume and total alveolar surface in fresh lung.

fraction $\varrho'_A = \varphi \cdot \varrho_A$ was obtained by volumetric analysis as described on p. 54 (Table V.2 and Fig. 34). The average coefficient of configuration $\bar{\sigma}_A$ relating surface and volume of one alveolus was estimated from models to be 4.8 (cf. p. 61); it is not very sensitive to slight variations in shape, so long as the basic configuration of an open poly- hedron is preserved. In addition, it could be shown secondarily to be 4.87 on the average (Table VI.1).

The total alveolar surface areas obtained by equation (3.93) in our five cases are listed in Table VI.4. In the fresh state S_{AT1} ranged from 43 to 80 m², depending mainly on the size of the lungs.

2. Estimation of Alveolar Surface Area by Mean-chord-length Method

The calculation of the alveolar surface from equation (3.95) proposed by TOMKEIEFF (1945) and HENNIG (1956) requires the measurement of the "mean chord length" L_m of the air spaces of the lung as outlined on p. 37. The values of L_m obtained on our five lungs are listed in Table VI.5. It can be seen that L_m increases with the size of the lung; measured on processed tissue it is approximately 200 μ in the adult lungs at the given degree of 3/4 total inflation. This dimension does not represent any particular alveolar dimension since it is the "mean chord length" of all parenchymal air spaces, including central air ducts. Introducing L_m and the corresponding parenchymal volume φV_{LS} into equation (3.95) we obtain the alveolar surface in the processed lung:

$$S'_{AT2} = 4 \cdot \frac{\varphi V_{LS}}{L_m} .$$

Table VI.5. *Alveolar Surface by Tomkeieff-Hennig Method (2)*

#	V_{LS} cm³	L_m 10^{-4} cm	S'_{AT} 10^4 cm²	S_{AT} 10^4 cm²
2471	1050	157	24.0	43.7
555	1090	148	26.4	48.0
9975	2340	188	44.6	81.2
8710	2890	249	41.8	76.1
3361	2420	218	40.0	72.9

V_{LS} = volume of processed lung,
L_m = mean linear intercept (compare text),
S'_{AT} = total alveolar surface in processed lung,
S_{AT} = total alveolar surface in fresh lung.

The values of S'_{AT2} computed for our cases are again listed in Table VI.5, as well as those of S_{AT2} which were obtained by applying the conversion factor $(1/f_p)^2$ to S'_{AT2} (cf. p. 45, Table IV.2). This second estimate of S_{AT} again appeared to increase with lung size. The individual values show satisfactory agreement with those obtained by the geometric method.

3. Total Alveolar Surface from Linear Dimensions of Alveoli

The linear dimensions of the simplified model alveolus illustrated in Fig. 41 d allowed us to calculate, in an approximate manner, its average volume $\bar{v}_A$ and surface $\bar{s}_A$ by equations (6.1) and (6.2). The total alveolar surface can now be obtained by

$$S_{AT3} = \frac{\varrho'_A V_L \cdot \bar{s}_A}{\bar{v}_A} \qquad (6.10)$$

Table VI.6. *Alveolar Volume and Surface Calculated from Linear Dimensions (Method 3)*

#	V_L cm³	$\bar{v}_A$ 10^{-6} cm³	$\bar{s}_A$ 10^{-4} cm²	S_{AT} 10^4 cm²
2471	2580	5.44	15.51	44.4
555	2680	5.34	15.52	46.8
9975	5750	12.53	27.45	75.8
8710	7100	17.90	31.15	74.2
3361	5950	14.91	30.65	73.4

V_L = volume of fresh lung,
$\bar{v}_A$ = volume of 1 average alveolus in fresh state,
$\bar{s}_A$ = surface of 1 average alveolus in fresh state,
S_{AT} = total alveolar surface in fresh lung.

$\varrho'_A V_L$ represents again the alveolar fraction of the total lung volume (cf. Fig. 34). Since the linear dimensions had been converted into the fresh state, the fresh lung volume V_L can be introduced directly. The values of $\bar{s}_A$, $\bar{v}_A$ and S_{AT} are listed in Table VI.6. Although these estimates of S_{AT3} can be considered only rough approximations, they show quite satisfactory agreement with the two other independent determinations of S_{AT}.

4. Discussion and Correlation of Findings

Three independent methods provided estimates of the alveolar surface area S_{AT} which showed close agreement in all five cases investigated. In adults S_{AT} was approximately 70 to 80 m². This result is consistent with an alveolar number of 300 million and agrees with the older and frequently quoted dimensions (Table I.1; cf. VON HA-YEK, 1953). It differs significantly, however, from more recent results obtained by KULENKAMPFF (1957) and HIERONYMI (1961) who found a surface area of 25 to 30 m². While in most of the older studies apparent differences may be partially ascribed to inadequate methods of measurement (see p. 2), KULENKAMPFF and HIERONYMI used the reliable method of "mean chord length" determination which was also employed here. The

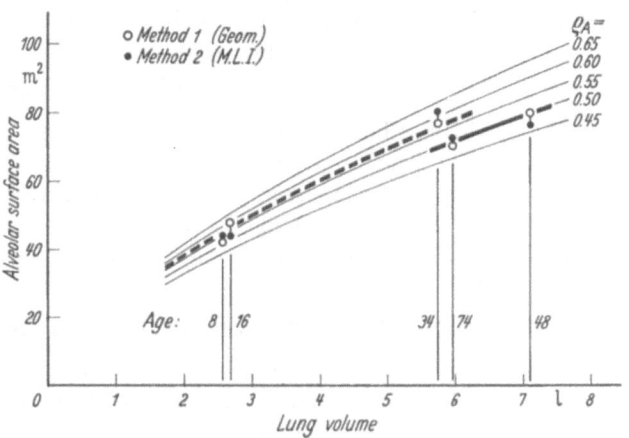

Fig. 47. Alveolar surface area as a function of the lung volume and its alveolar fraction ϱ_A. Since ϱ_A was found to decrease with age, S_{AT} was relatively smaller in the older cases (heavy curve) than in the younger (broken line). Values of S_{AT} obtained by geometric and mean chord length method are plotted as open and full circles, respectively
[From Weibel, 1962]

investigation of lungs in different degrees of inflation may have caused some of the variability of estimates; but this is not sufficient to explain the low estimates of these two recent studies, since an alveolar surface area of 25 m² is entirely consistent with the finding of a total number of only 70 million alveoli as obtained by ELZE and HENNIG (1956) on the basis of KULENKAMPFF's data and by HIERONYMI. The possible causes of this significant underestimation of the alveolar surface area were discussed on p. 63 in reference to the alveolar number, which should not depend on the degree of inflation at which the lungs were examined. The arguments brought forth led us to the conclusion that these two studies could not invalidate our results because they may have included two major sources of error which were carefully avoided in the present investigation. It should also be noted that they share possible errors since HIERONYMI (1961) adopted KULENKAMPFF's (1957) and HENNIG's (1956) procedure, including some numerical constants, without criticism. Therefore, we remain confident in our results.

In Fig. 47 the alveolar surface area is related to the lung volume V_L. The values of S_{AT1} and S_{AT2} obtained by the geometric and "mean-chord-length" methods, respectively, are plotted as open and closed circles. The value S_{AT3} is not included in this argument despite excellent agreement; it was not obtained by random measurements and, therefore, does not carry undue weight. Equation (3.93) represents the formulation of the surface-to-volume relationship of a honeycomb-like geometric analog of the respiratory zone proposed above (Fig. 37 and 46). A fraction ϱ'_A of the lung volume V_L is occupied by a number N_{AT} of alveoli which have an average surface-to-volume ratio $\bar{\sigma}_A$, so that its total internal surface is

$$S_{AT} = N_{AT} \cdot \bar{\sigma}_A \left(\frac{\varrho'_A \cdot V_L}{N_{AT}} \right)^{2/3}. \tag{3.93}$$

Of these factors N_{AT} and $\bar{\sigma}_A$ have proved essentially constant (Fig. 42 and Table VI.1). The volume V_L varied from case to case. The volumetric fraction ϱ'_A showed a significant reduction occurring with age, so that ϱ'_A should be regarded as a variable. We, therefore, find that

$$S_{AT} = 3.22 \cdot 10^3 \cdot (\varrho'_A \cdot V_L)^{2/3}. \tag{6.11}$$

In Fig. 47 this has been plotted parametrically in ϱ'_A. In an attempt to reconcile our data to this family of curves, the broken line indicates the isopleth of $\varrho'_A = 0.56$ which corresponds to the average of ϱ'_A in the three younger cases; the solid line refers to an average $\varrho'_A = 0.50$ observed in the older lungs (Fig. 34). It is evident that these functions correspond to the observed values of S_{AT} rather satisfactorily.

E. Relation of Estimated Alveolar Surface to Actual Air-tissue-interface

The alveolar surface area S_{AT} estimated in the preceding paragraphs is proportional but not equal to the extent of the air-tissue-interface I_{AT}. From the point of view of gas exchange I_{AT} must be regarded as the significant quantity, so that we need to estimate the magnitude of the proportionality coefficient γ of the relationship.

$$I_{AT} = \gamma \cdot S_{AT}. \tag{6.12}$$

In general we can state that $\gamma > 1$ since S_{AT} must represent an underestimate of I_{AT}. The diagram of Fig. 48a shows a section of an interalveolar septum (cf. Fig. 66). The low magnifications and the section thickness of $7\,\mu$ used in analyzing S_{AT} did not allow us to resolve the surface texture of the interalveolar septa. Rather, we have measured the area of a smooth surface (Fig. 48b). The alveolar epithelium, however, follows the outline of the capillary network to a certain extent. These capillaries seem to "bulge" slightly into the alveolus so that a small fovea is formed in the center of each capillary mesh (Fig. 48c). This may cause the air-tissue-interface to be about 20% larger than the "alveolar surface". An additional increase of this interface area may result from fine crenations of the alveolar epithelial cells. These crenations may, however, be partially smoothed out by the fluid covering the alveolar wall (MACKLIN, 1956; CHASE, 1959) so that their quantitative effect is not estimated here. Therefore, we arrive at the conclusion that it may be appropriate to let $\gamma = 1.2$ so that

$$I_{AT} = 1.2 \cdot S_{AT}$$

represents a satisfactory estimate of the extent of the air-tissue-interface.

F. Effect of Constitutional Lung Size and Different Degrees of Inflation on Alveolar Dimensions

In this study the size of the lung is represented by the lung volume V_L which was shown for all cases to be approximately 3/4 of the hypothetic maximal lung volume V_{TLC}. The actual lung volume V_L is therefore a function of the degree of

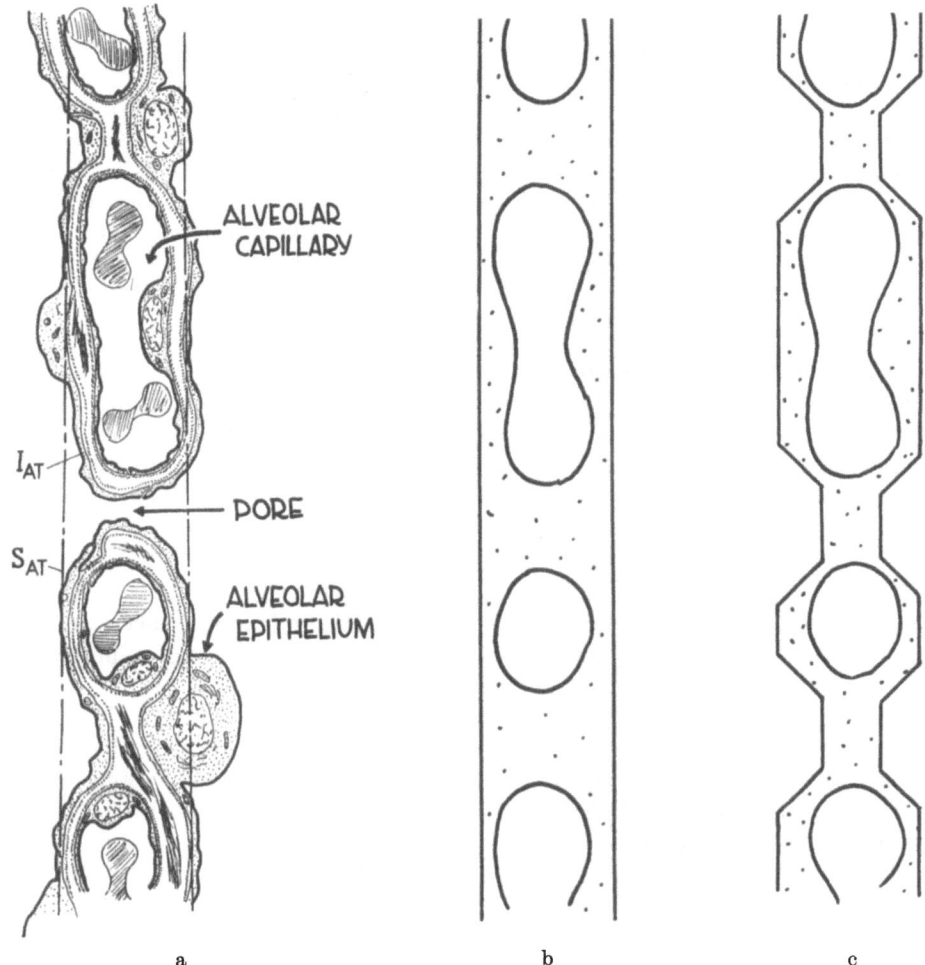

Fig. 48. Relation of alveolar surface S_{AT} to actual air-tissue interface I_{AT}. S_{AT} is measured as flat surface while I_{AT} is somewhat corrugated. Compare text

inflation. We accept the hypothesis (cf. p. 41) that the anatomic volume of the lung at "total lung capacity" V_{TLC} is representative of the constitutional lung size and provides a baseline value which permits comparison of individual cases. The total lung capacity of normal individuals has been shown repeatedly to be related to the body size. BALDWIN et al. (1948) related it to the sitting height of their patients, BARNHARD et al. (1960) to the body surface area. TENNEY and REMMERS (1962), on the other hand, observed, in a comparative study, that the maximal lung volume of various species was proportional to the body weight. Since the relevant information was incomplete, we have not attempted such correlation for our few cases. The gross variation in volume observed in these five lungs is easily explained by the increase in body size occurring during growth.

The question now is, how does the over-all size of the human lung influence the dimensions of alveoli. Although this point has rarely been considered, recent investigations of "alveolar dimensions" on the basis of surface tension phenomena by CLEMENTS (1961) seemed to suggest that the number of alveoli increases with the constitutional lung size (weight!) while the alveolar "mean radius" does not appear to vary greatly. This is in contradiction to our findings. We have demonstrated that all our five lungs contained a constant number of about 300 million alveoli despite an almost threefold variation in lung volume. Correspondingly, it was found that the dimensions of single alveoli depend on the over-all size of the lung. At the investigated degree of inflation ($V_L = .75\ V_{TLC}$) the "mean diameter" d_A of the alveoli followed quite closely the function

$$d_A = 1.54 \cdot 10^{-3} \cdot V_L^{1/3}$$

as can be seen in Fig. 45. Evidently, d_A must also be influenced by the degree of inflation of the lungs. On p. 44 it was suggested that the lung volume V_L was a combined function of the constitutional lung size V_{TLC} at maximal inspiration and of the fractional degree of inflation ζ. A more general form of equation (6.9) will, therefore, relate the "mean diameter" d_A to the constitutional lung size V_{TLC} as well as to the degree of inflation ζ:

$$d_A = 1.54 \cdot 10^{-3} (\zeta \cdot V_{TLC})^{1/3} . \tag{6.13}$$

This was taken into consideration in Fig. 45 by plotting a family of curves around the abscissa which indicates, along the isopleths, the change in lung volume with varying degrees of inflation ζ.

The "mean diameter" of an alveolus is, actually, a quantity with little geometric meaning; it is merely indicative of alveolar dimensions. It may be of more significance to consider the relation of the mean alveolar volume $\bar{v}_A$ and of the mean alveolar surface $\bar{s}_A$ to constitutional lung size and degree of inflation. In a general fashion we obtain

$$\bar{v}_A (\zeta, \mathrm{TLC}) = \frac{\varrho'_A(\zeta) \cdot \zeta \cdot V_{TLC}}{N_{AT}} \tag{6.14}$$

$$\bar{s}_A (\zeta, \mathrm{TLC}) = \bar{\sigma}_A(\zeta) \cdot \left[\frac{\varrho'_A(\zeta) \cdot \zeta \cdot V_{TLC}}{N_{AT}} \right]^{2/3} \tag{6.15}$$

We observe that the degree of inflation can influence both dimensions by reducing first the lung volume through the variable factor ζ. In addition, it is quite likely that the volumetric fraction of alveoli ϱ'_A is also influenced, but we do not know the nature of the function $\varrho'_A(\zeta)$. This important point needs experimental investigation. It is to be anticipated that the surface-to-volume coefficient $\bar{\sigma}_A$ is also a function of ζ. The alveoli, however, must change their shape quite drastically to significantly influence $\bar{\sigma}_A$ (cf. p. 37, Fig. 23).

The mean alveolar diameter d_A was defined as the diameter of the spheric analog of the alveolus having equal volume. Its most general equation follows, therefore, from (6.13):

$$d_A(\zeta, \text{TLC}) = \left[\frac{6 \cdot \varrho'_A(\zeta) \cdot \zeta \cdot V_{TLC}}{\pi \cdot N_{AT}}\right]^{1/3} \tag{6.16}$$

In equation (6.13) we have evidently neglected the effect of a change in ϱ'_A with varying ζ which, in view of the approximate meaning of d_A, we are justified in doing since this change may be small within the range of $0.5 < \zeta < 1.0$ considered. Such changes would, however, not be negligible with respect to $\bar{v}_A$ and $\bar{s}_A$.

The last and most significant question concerns the effect of constitutional lung size and degree of inflation on the total alveolar surface area S_{AT}, or, still more important, on the extent of the air-tissue-interface I_{AT}. On the basis of equations (6.15) and (3.93) we arrive at the general expression for S_{AT}:

$$S_{AT}(\zeta, \text{TLC}) = N_{AT} \cdot \bar{\sigma}_A(\zeta) \cdot \left[\frac{\varrho'_A(\zeta) \cdot \zeta \cdot V_{TLC}}{N_{AT}}\right]^{2/3}. \tag{6.17}$$

Depending on the unknown nature of the functions $\varrho'_A(\zeta)$ and $\bar{\sigma}_A(\zeta)$, the alveolar surface area may change with varying degrees of inflation (or inspiration). ELZE and HENNIG (1956) considered, theoretically, the effect on S_{AT} of geometric deformations of a simple cuboid alveolar model as may be assumed in the ventilating lung. They arrived at the conclusion that the alveolar surface of an adult may vary from 34 to 72 m² between maximum expiration and inspiration. In quiet breathing they expect S_{AT} to vary from 51 to 56 m². Their reasoning, however, was based on little evidence so that these figures should be considered suggestive but not conclusive.

The physiologically important quantity is the extent of the air-tissue-interface I_{AT} which is related to S_{AT} through the proportionality coefficient γ [p. 70, equation (6.12)]. It is to be expected that $\gamma = \gamma(\zeta)$ is also dependent on the degree of inflation, so that we find

$$I_{AT}(\zeta, \text{TLC}) = \gamma(\zeta) \cdot S_{AT}(\zeta, \text{TLC}). \tag{6.18}$$

In estimating the trend of $\gamma(\zeta)$ we observe that it relates to the surface texture of the alveolar wall which is disregarded in estimating S_{AT}. This surface texture may become partially flattened out during deep inflation, so that $\gamma(\zeta)$ would be inversely proportional to ζ. In conclusion it follows, therefore, that, although S_{AT} may vary with ζ, the actual air-tissue-interface may change but little with different degrees of inflation, at least within a somewhat physiologic range. These points call for experimental investigation.

Chapter VII

Geometry and Dimensions of Alveolar Capillary Network

A. Architecture of the Alveolar Capillary Network

The basic element of the *systemic* capillaries is, in the vast majority of instances, a rather long thin tube, which often is connected at either end with other capillaries. Thus there is built up a capillary network comprising more or less loose meshes. The *pulmonary* or *alveolar* capillaries, by contrast, form a dense network which is enclosed in the alveolar wall. This network assumes its simplest two-dimensional form within each single facet of the interalveolar septa (Figs. 49 and 50). Its basic elements, the capillary segments, fundamentally have the shape of short cylindric tubes, as shown in Fig. 51, these being modified at their bases in the form of

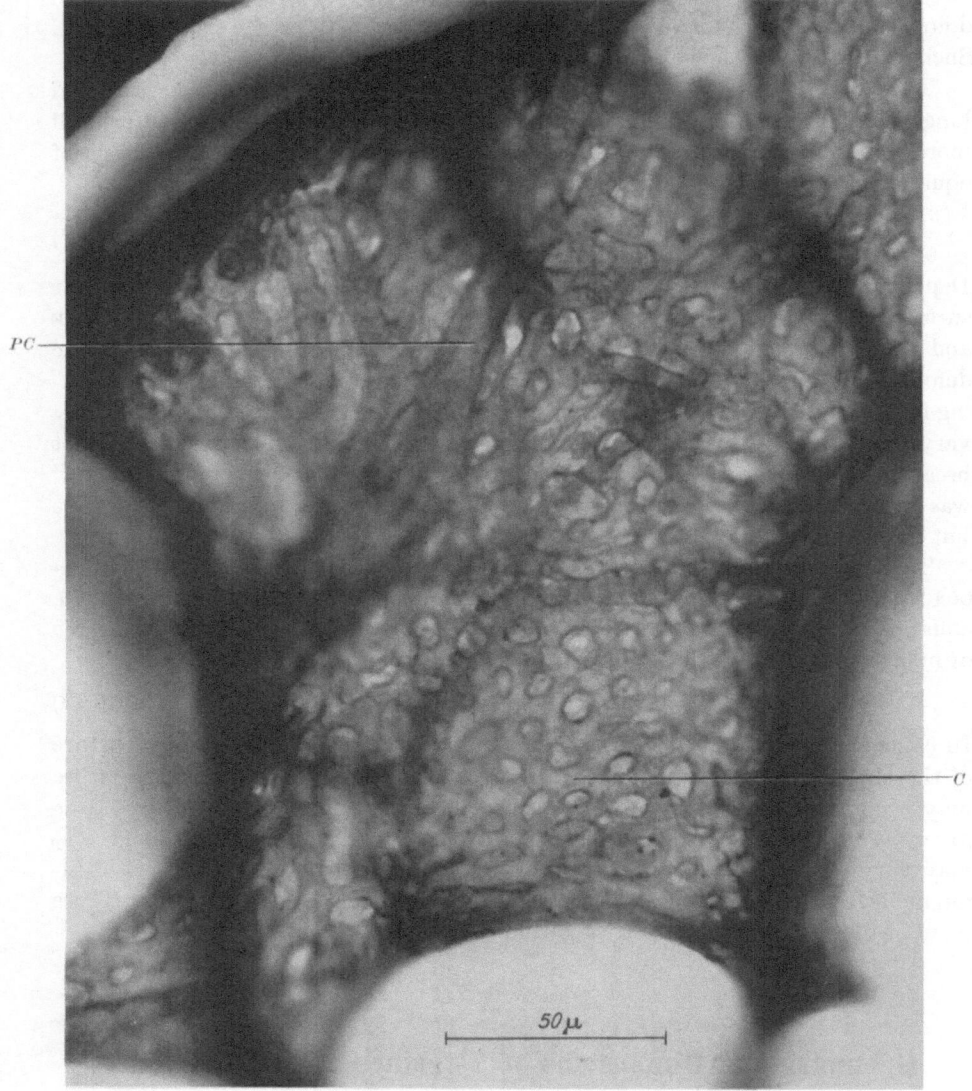

Fig. 49. Flat aspect of interalveolar septum showing capillary network (C) by staining of the capillary basement membranes by PAS reaction. Fan-shaped precapillary (PC) leads into network. Human lung. × 580 [From WEIBEL and GOMEZ, 1962b]

wedges, in order to allow each segment to join at either end with two adjacent segments. If we examine the points of junction of a network, as depicted in Figs. 49 and 50, we observe that, in the vast majority of instances, three segments join, but junctures of four may occasionally occur. This results in an average angle of juncture of about 120°, although the individual angle may deviate from this average. As a further consequence of this arrangement the basic geometric structure of this capillary network must be hexagonal. This means that each mesh of the capillary network must, on average, be surrounded by six segments. This arrangement is, of course, only true in a schematic sense, since size and shape of the meshes and segments in the true network are decidedly irregular (Fig. 49). However, if we analyze a large number of tracings of capillary networks, as illustrated

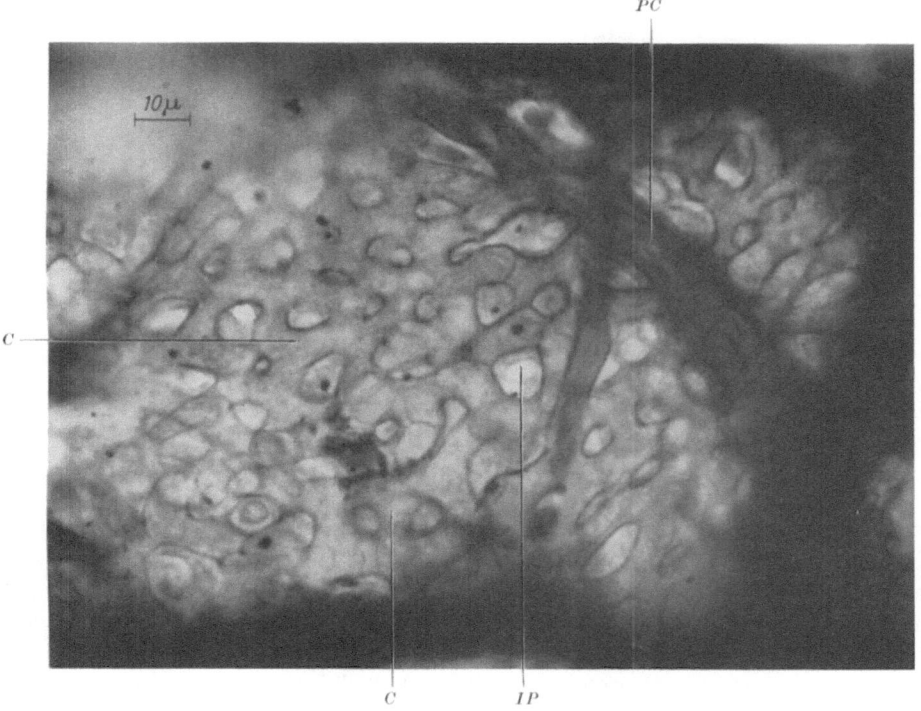

Fig. 50. Flat aspect of capillary network (C) in interalveolar septum of human lung at higher magnification Pre- or postcapillary with lateral connections to network (PC) and interalveolar pore (IP) are seen. × 750
[From Weibel, 1962]

in Fig. 52, we shall find mostly 5, 6, or 7 segments in a mesh, with an average of almost exactly 6, so that the fundamental concept of a hexagonal network is supportable, despite the admitted irregularities.

This two-dimensional network becomes three-dimensional where three septal facets join, as described on p. 77 (Fig. 53). It is achieved, in principle, by rotation of the plane of juncture of 3 segments by approximately 90°. Because of this rotation, each of the segments of this tripod lies in a different septal facet and can, accordingly, achieve connection with the network within its facet (Fig. 53). As a result of this arrangement, the alveolar capillary network is continuous throughout many interalveolar facets. At present, it is difficult to estimate how far this continuity extends. It appears, however, that it may certainly continue throughout pulmonary units such as lobules. In human lungs, with their incomplete separation of structural units, it may well be partially continuous throughout segments, or even throughout lobes. This is suggested by the observation of collateral ventilation across pulmonary segments (VAN ALLEN et al. 1930, LIEBOW 1959) which indicates continuity of pulmonary parenchyma and therefore renders continuity of capillaries possible. Whether this continuity would have functional significance remains uncertain.

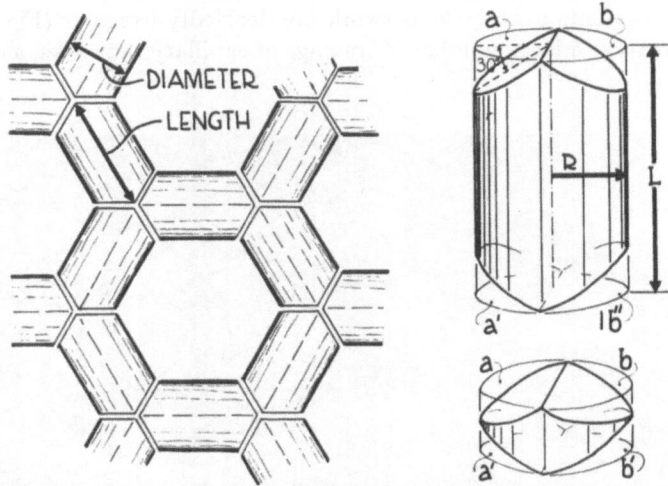

Fig. 51. Geometric model of capillary network. Models of capillary segments used for computation of volume and surface. Compare text [From Weibel, 1962]

B. Geometry of Capillary Segments

In the simplest approximation the elements of the capillary network can be described as short cylinders (Fig. 51 b). At either end of these segments two wedges (a, b and a', b') are cut off at an angle of 30° to the base. The two cut surfaces enclose an angle of 120° which we have described as the average angle of segmental juncture in a hexagonal network. It allows three of these "wedged" cylinders to be fitted together end to end (Fig. 51 a). This simplified model segment is defined by two dimensions:

L_c: its axial length from intersection to intersection,
and
R_c: the radius of the cylinder.

From the shape already described, we can derive the volume v_c of the segment, and its cylindric surface s_c (without cut surfaces) as being:

$$v_c = R_c^2 (\pi L_c - 1.54 R_c) \tag{7.1}$$
$$s_c = 2 R_c (\pi L_c - 2.3 R_c) . \tag{7.2}$$

If

$$L_c \geqq \frac{2}{\sqrt{3}} R_c . \tag{7.3}$$

If L_c is smaller, the wedges a, b will overlap a', b', resulting in an underestimation of volume or surface if the above coefficients are employed. For the case where $L_c \leqq \left(\frac{2}{\sqrt{3}}\right) R_c$ equations (1 and 2) have been modified in order to avoid this underestimation. We then approximate volume and surface by

$$v'_c = \left(\pi - \frac{4}{3}\right) L_c \cdot R_c^2 \qquad (7.4)$$

$$s'_c = 2(\pi - 2) L_c \cdot R_c . \qquad (7.5)$$

Geometrically, this means that the wedges can touch at their base (Fig. 51 c).

The assumption of a cylindric shape for the capillary segment is an oversimplification of the true situation. Figure 54 demonstrates graphically that the junction of two segments is not characterized by a sharp break in the outline of the two capillary segments as is implied in the model of Fig. 51. Rather, the basement membranes of the two segments show a smoothly curved transition from one to the other. This can easily be observed in Figs. 49 and 50. As a result of this, the cross section of the capillary will be circular only in the central portion of the segment, while it will assume a somewhat elliptic shape toward the extremities. Since the

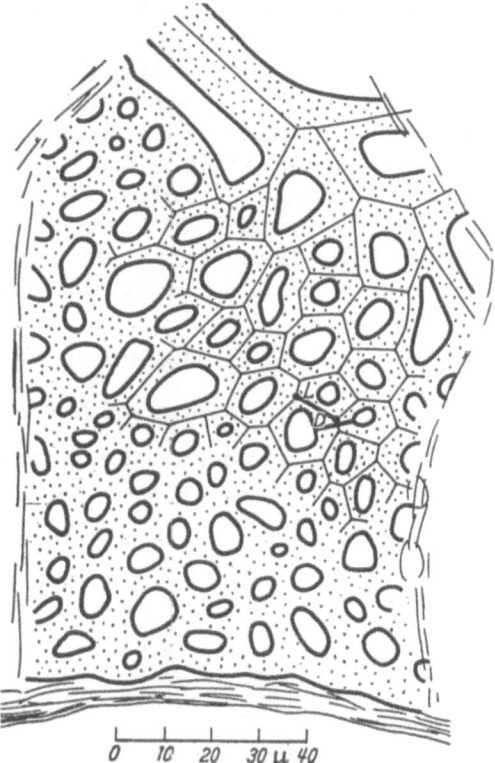

Fig. 52. Tracing of capillary network as used for measurement of dimensions of capillary segments [From WEIBEL, 1962]

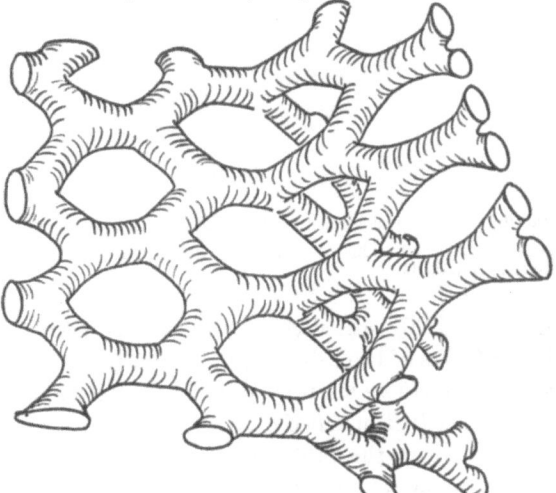

Fig. 53. Schematic sketch of capillary network at juncture of three interalveolar septal facets showing continuity of network

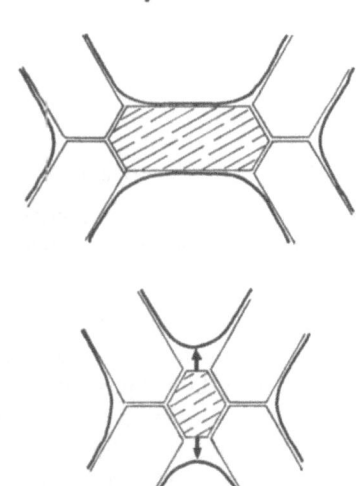

Fig. 54. Effect of smooth contour of capillary basement membrane on apparent diameter of short segments

radius R_c represents the smallest radius of the capillary segment in its central portion (see below), the volumes and surfaces calculated by the above equations will prove to be slight underestimates of the actual dimensions. If the capillary segmentis very short (Fig. 54b), the smoothly curved transition of the basement membrane will tend to "pull over" from the segment joining the short one on one side to the segment on the other side. This phenomenon has actually been observed. It causes the diameter of short segments to be relatively larger than that of long segments (see Fig. 57), as will be discussed. In these circumstances the capillary cross-section will never be circular but always elliptic. Because we are measuring the larger semi-axis as the radius R_c we shall, thus, overestimate volume and surface of these segments. The geometric simplifications introduced have caused overestimations of volume and surface on the one hand and underestimations on the other. However, these errors are small; since they will, on the whole, tend to cancel each other out, we can ignore them.

C. Dimensions of the Capillary Network

1. Dimensions and Number of Capillary Segments

The capillary network was demonstrated by staining its basement membrane with the PAS-reaction (p. 47). Intense light absorption in the lateral portions of

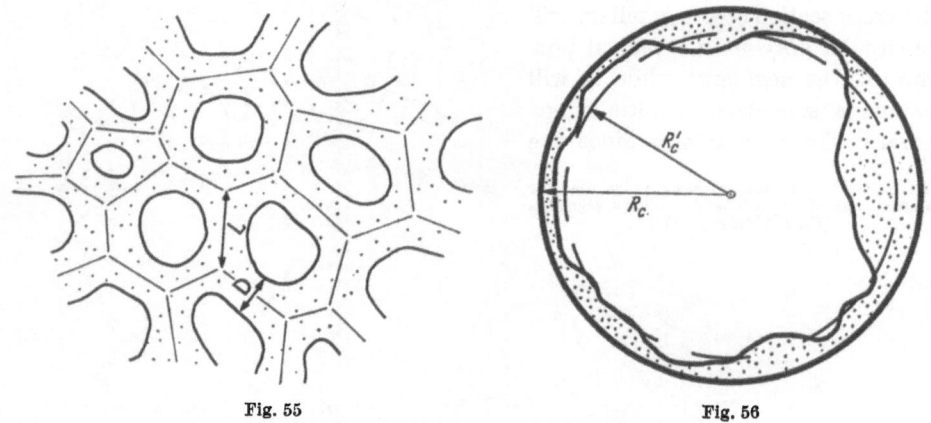

Fig. 55   Fig. 56

Fig. 55. Fragment of tracing of capillary network with indication of measurement of dimensions of segments

Fig. 56. Relation of external radius R_c to average internal radius R'_c of capillary. Stippled area represents endothelium

this basement membrane sleeve (Figs. 28 and 29) caused the capillary network to be sharply outlined in whole unsectioned facets of the interalveolar septa (Figs. 49 and 50). For the purpose of measurement, facets which were parallel to the plane of section were sampled according to the procedure outlined on p. 16. The capillary network enclosed in this facet was projected on the table at a magnification of 1200:1 through the Wild camera lucida attachment, and the contours of the network were traced. This produced, for each case, a set of tracings, an example of which is shown in Fig. 52. On these tracings the central axes of the individual capillary segments were located and drawn in as straight lines (Fig. 55). The intersection of axes marked the axial point of segmental junction. The length L_c was measured along the axis as the distance between the two axial junction points. The diameter $D_c = 2R_c$ was measured perpendicularly to the axis as the shortest distance between the opposite outlines of the segment.

The diameter thus obtained qualifies as the external capillary diameter (see Figs. 29 and 56). It corresponds to the lumenal diameter plus the double thickness of the capillary endothelium. As shown in the diagram of Fig. 56, the lumenal radius R'_c follows from R_c by subtracting the average thickness of the endothelium $\bar{\tau}_e$:

$$R'_c = R_c - \bar{\tau}_e \,. \tag{7.6}$$

As will be discussed on p. 103 $\bar{\tau}_e$ is estimated to be about 0.3 μ.

The number of capillary segments per unit area was counted on rectangular fields superimposed on the tracings. This yielded the average number n_c of capillary segments per unit area of interalveolar septum. The total area of the interalveolar septum S_{AM} could be approximated by

$$S_{AM} = \frac{1}{2} \cdot S_{AT} \tag{7.7}$$

since the alveolar surface area S_{AT} represents an estimate of the surface of the double-faced interalveolar septa. The total number of capillary segments in the entire lung is therefore

$$N_{CT} = S_{AM} \cdot n_c \,. \tag{7.8}$$

In each of our five lungs (Table IV.1) a total of 600 to 1,500 capillary segments were measured on about 20 tracings. Figure 57 presents the measurements obtained in two cases: The individual diameters D_c are plotted against the corresponding lengths L_c, each dot representing one segment. The dimensions are widely scattered in all directions. This is also apparent from the distribution curves of D_c and L_c shown in Fig. 58; the diameters spread over a narrower range than did the lengths but still they varied between 1 and 15 μ. In Fig. 57 a correlation between diameter and length is apparent, the general trend being "the shorter the wider". Upon closer examination, however, one observes that this trend is present only in the short range, where $L_c < 8\ \mu$, while no correlation between L_c and D_c is apparent with longer segments. This peculiarity can be explained by the behavior of the capillary basement membrane which was shown above (Fig. 54) to smooth out the hypothetic corners of our model network and to widen the lateral diameter of shorter segments by "pulling over" between two adjacent segments (Fig. 54b).

The frequency distribution curves of D_c and L_c in the five investigated cases have been superimposed in Fig. 59. The distribution of diameters shows a sharp peak which coincides in three of the cases, while it is shifted toward smaller values in Case 555 and toward larger diameters in Case 8710. The average external diameters of the capillary segments ranged from 5.1 to 7.7 μ in the processed state, with an average of 6.6 μ. In reference to the fresh lung, the external diameter averages 8.6 μ; allowing 0.3 μ for the average thickness of the capillary endothelium (cf. p. 103, Figs. 56 and 75), the average internal capillary diameter is estimated to be 8 μ. The individual average capillary diameters in the five cases investigated are listed in Table VII.1. We find that the average capillary diameter must be related to the degree of capillary filling. The smallest value of 6.1 μ was found in Case 555 in which there had been severe blood loss due to splenic rupture just prior to death. The small capillary diameter, therefore, reflects a low degree of vascular filling. The largest values of D_c (8.6 μ and 9.6 μ) were observed in

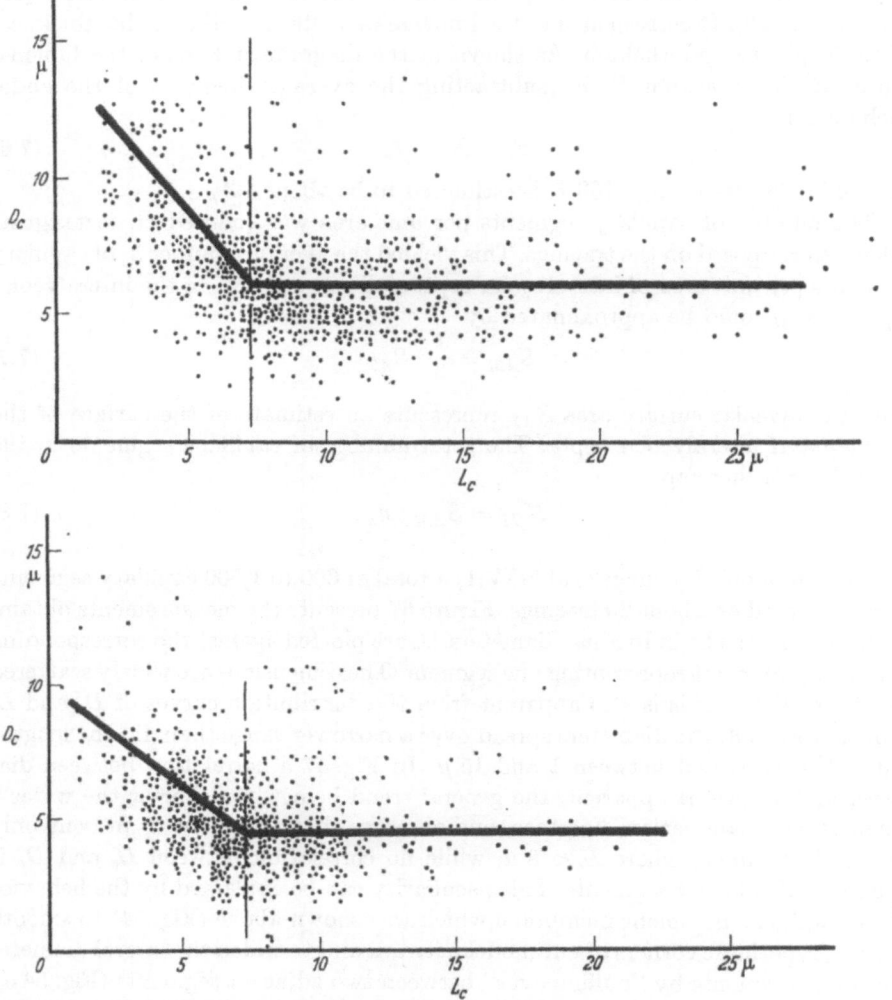

Fig. 57. Dispersion diagram of diameter D_c and length L_c of capillary segments in two cases. Each dot represents one segment. Dimensions refer to processed lung

lungs which also showed a mild degree of pulmonary edema, Cases 9975 and 8710, thus indicating the presence of pulmonary vascular congestion (cf. Table IV.1).

The average values of the axial length of the capillary segments are also listed in Table VII.1. They ranged from 9 to 13 μ after conversion of the measurements into values for the fresh state. It appears that L_c is proportional to the size of the lung, therefore it seems most appropriate to relate L_c to the square root of the alveolar membrane area S_{AM} on which the capillaries are spread as a two-dimensional network (Figs. 49 to 51). By "least square" fitting it is, indeed, found that a function

$$L_c = 2.14 \cdot 10^{-6} \sqrt{S_{AM}} \qquad (7.9)$$

well describes the findings, as shown in Fig. 60a.

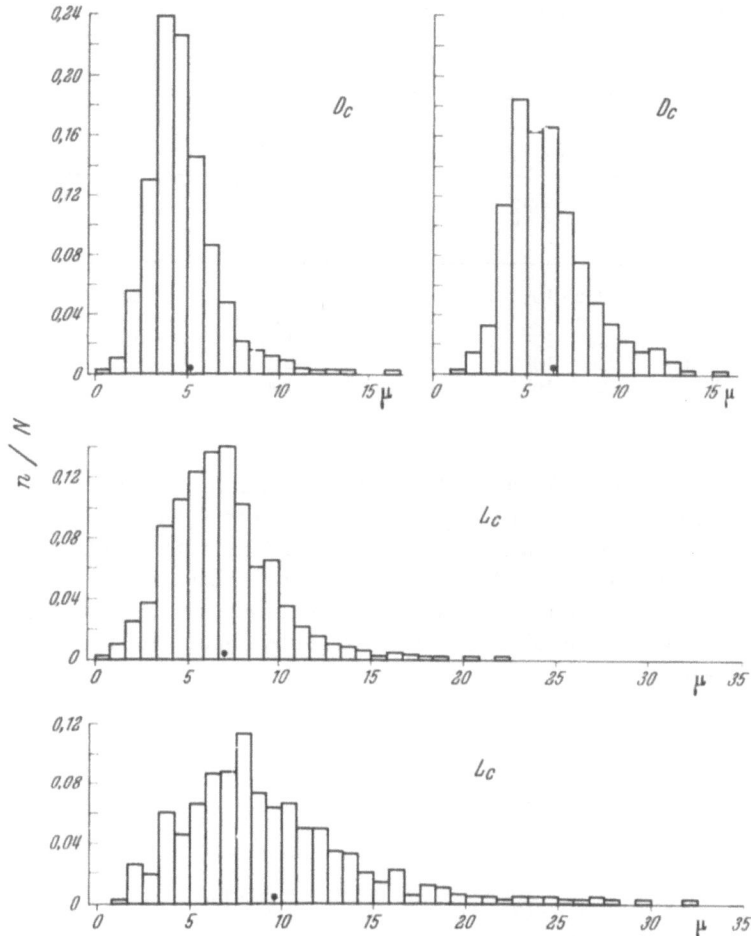

Fig. 58. Distribution of diameter and length of capillary segments in two cases. Dimensions refer to processed lung
[From Weibel, 1962]

On the other hand, it is evident that the average number of capillary segments per 1 cm² of membrane area S_{AM} is inversely proportional to the lung size. The average numbers and the standard deviation of the counts in our five cases are again listed in Table VII.1; the averages are plotted in Fig. 60b against S_{AM}. If the total number of capillary segments in the entire lung is calculated by equation (7.8), it is found to be independent of lung size (Fig. 60b, open circles). The group average is about 280 billion, with only small individual variations, as can be seen in Table VII.2.

Table VII.1. *Dimensions and Number of Alveolar Capillary Segments*

#	Average length[1] $\bar{L}_C$		Average internal diameter[1] $\bar{D}_C$		Number of capillary segment per 1 cm² alveolar membrane[2]	
	10^{-4} cm	S. D.	10^{-4} cm	S. D.	n_C	S. D.
2471	10.7	± 4.5	8.3	± 3.1	$212 \cdot 10^4$	$\pm 28 \cdot 10^4$
555	9.5	± 3.9	6.3	± 2.4	$237 \cdot 10^4$	$\pm 37 \cdot 10^4$
9975	12.4	± 4.0	9.0	± 2.9	$142 \cdot 10^4$	$\pm 23 \cdot 10^4$
8710	14.2	± 5.2	9.9	± 3.5	$118 \cdot 10^4$	$\pm 24 \cdot 10^4$
3361	13.1	± 6.0	8.3	± 2.9	$141 \cdot 10^4$	$\pm 36 \cdot 10^4$

[1] Refers to fresh lung. — [2] Refers to processed tissue.

Table VII.2. *Over-all Dimensions of Capillary Network*

#	Area of inter-alveolar septa[1] S_{AM} 10^4 cm²	Number of capillary segments N_{CT}	Total capillary volume[1] V_{CT} ml	Estimated blood volume[1] V_{CB} ml	Total capillary surface[1] S_{CT} 10^4 cm²	Surface of blood[1] S_{CB} 10^4 cm²
2471	21.8	$252 \cdot 10^9$	124	108	48.0	44.0
555	23.5	$298 \cdot 10^9$	87	75	43.5	40.2
9975	39.5	$302 \cdot 10^9$	195	168	73.0	67.7
8710	39.0	$260 \cdot 10^9$	244	212	82.5	76.5
3361	36.0	$275 \cdot 10^9$	192	166	70.2	65.1

[1] Refers to fresh lung.

We found earlier (p. 62, Table VI.2) that the number of 300 million alveoli building up the lung was a fixed property of this organ which appeared not to vary with lung size. We now find, in addition, that the total number of capillary segments building up the alveolar capillary network is also constant and independent of lung size. Since the capillary network is in contact with two alveoli, we deduce from this that each alveolus is surrounded by about 1,800 to 2,000 capillary (half-)segments.

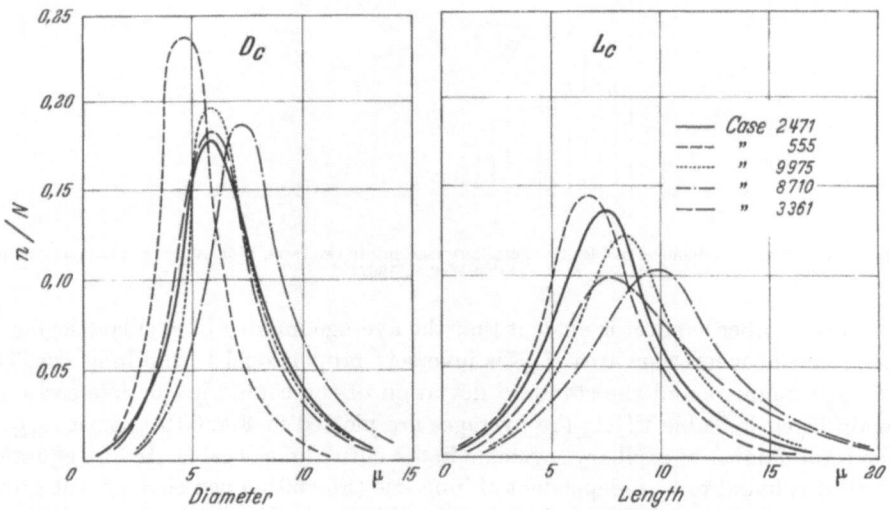

Fig. 59. Superimposed distribution curves of diameter and length of capillary segments as found in the five human lungs investigated. Dimensions refer to processed lung [From WEIBEL, 1962]

2. Over-all Volume and Surface of the Alveolar Capillary Bed

Having estimated the dimensions and the total number of capillary segments in the human lung, we can now proceed to estimate the anatomic pulmonary capillary volume and the over-all surface of the capillary bed. Equations (7.1) to (7.5) were derived to enable us to calculate volume and surface of the individual segments from axial length L_c and diameter $D_c = 2 R_c$. The total volume V_{CT} and total surface S_{CT} of the capillary bed follow from

$$V_{CT} = N_{CT} \cdot \bar{v}_c, \qquad (7.10)$$

$$S_{CT} = N_{CT} \cdot \bar{s}_c. \qquad (7.11)$$

Therefore, the average volume $\bar{v}_c$ and average surface $\bar{s}_c$ of the individual segments in each case remain to be estimated. The wide range of L_c and R_c (Figs. 57 and 58) does not allow us simply to compute $\bar{v}_c$ and $\bar{s}_c$ from the average values of L_c and R_c; it is necessary to average individually computed volumes and surfaces. Due to the large number of segments measured in each case (~ 1000), this computation would have been highly laborious. It could, however, efficiently and reliably be carried out by the aid of an IBM 650 automatic digital computer[1], which was programmed according to the flow chart shown in Fig. 61[2]. The entry cards contained radius, length and corresponding frequency m; the computer

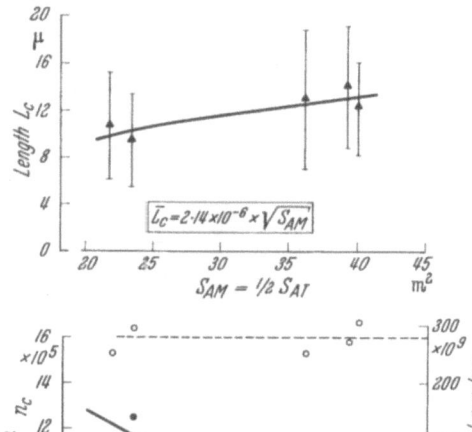

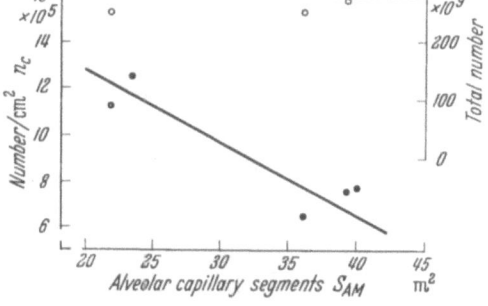

Fig. 60. a Relation of average length of capillary segments (triangles) to total area of interalveolar septa S_{AM}. Bars indicate standard deviation. b Number of segments per 1 cm² interalveolar septum in processed state (closed circles), and total number of capillary segments in entire lung (open circles) as function of S_{AM} [From WEIBEL, 1962]

used either equations (7.1) and (7.2) or (7.4) and (7.5) for computing v_c and s_c, multiplied them by m and computed the sums of v_c and s_c and m. The terminal card was programmed to initiate the computation of average and over-all volume and surface.

In Table VII.2 the values V_{CTS} and S_{CTS} refer to volume and surface of the capillary bed in the processed lung to which the corresponding correction factors of Table IV.2 were applied in order to obtain the values V_{CT} and S_{CT}, which represent the capillary volume and surface of the fresh lung. As already pointed out, the capillary radius R_c introduced into equations (7.1) to (7.5) represents the external capillary radius. The internal radius R'_c could be obtained by subtracting

---

[1] The author wishes to acknowledge his indebtedness toward the Watson Scientific Computing Laboratory of Columbia University for providing, under project 459, the necessary facilities and generous assistance.
[2] Upon request a duplicate of this program, for use on any 650 computer, may be obtained from the author.

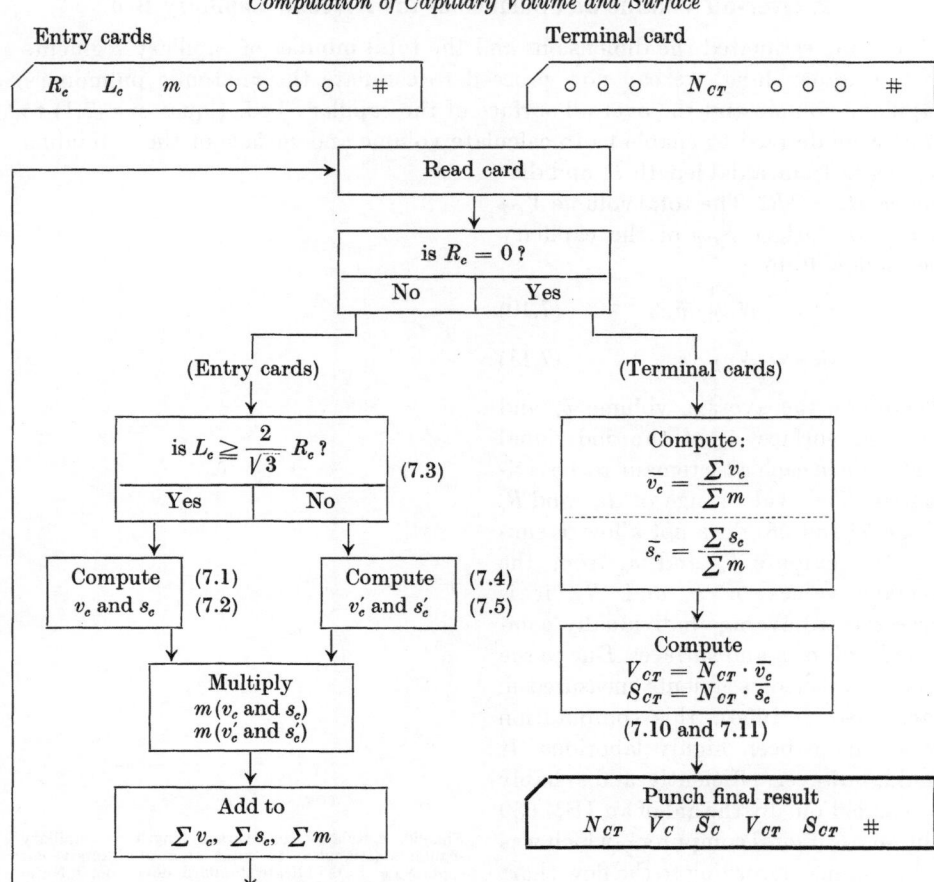

Fig. 61. Flow chart for computation of capillary volume and surface of entire lung with automatic computer

the thickness of the endothelium from R_c. This thickness was estimated to be approximately 0.3 μ (see p. 103). It follows, therefore, from geometric considerations that the capillary blood volume V_{CB} is roughly 85% of V_{CT}; it was found to be in the range of 75 to 210 cm³ in our five lungs. The internal capillary surface S_{CB} was estimated on the same basis to be about 92% of S_{CT}, being approximately 42 to 82 m² in our material.

Both S_{CB} and V_{CB} were found to be directly proportional to the size of the lung. This was indicated in Fig. 62 for the capillary surface which was plotted against the total alveolar surface, and in Fig. 63 for the capillary volume, plotted against the total lung volume. It is evident that the capillary surface S_{CB} is of the same order of magnitude as the alveolar surface, so that the air-tissue and the tissue blood interfaces are almost identical in their expanse (cf. p. 109).

It is obvious that the volume of the capillary bed and its surface will depend closely on the degree of blood filling. That the average capillary diameter did not appear to be related to the size of the lung was discussed above. However, the one case in which severe blood loss preceded death (Case 555, Table IV.1) was characterized by having an average capillary diameter clearly below that of the others, while the two showing a slight degree of pulmonary edema and congestion (Cases 9975 and 8710) had also the largest average capillary diameter. The average capillary diameter is, therefore, assumed to be proportional to the degree of capillary blood filling. The isopleths of Figs. 62 and 63 represent

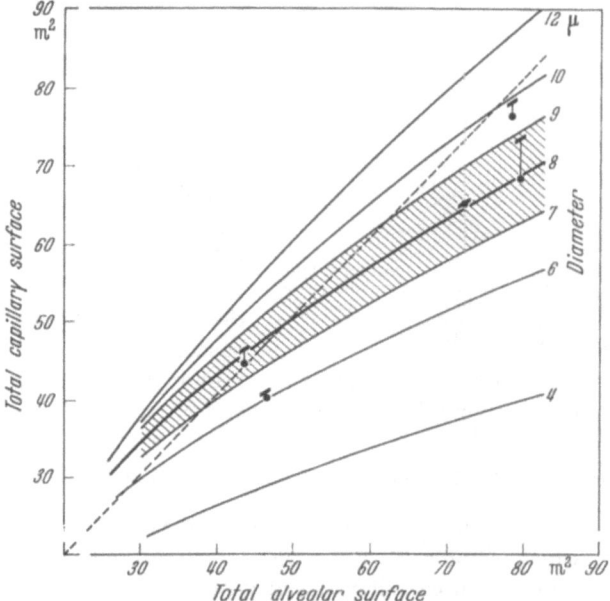

Fig. 62. Total internal capillary surface as function of the total alveolar surface (lung size) and of the average capillary diameter (degree of vascular injection). Full circles indicate the dimensions computed from the data; the horizontal bars refer the observed average diameter to the corresponding isopleths. Shaded area represents hypothetic "normal" range of variation of average diameter [From Weibel, 1962]

an attempt to express the changes in capillary surface and volume with the lung size at given average capillary diameters. These families curves express the equations:

$$S_{CB} = N_{CT}[9.48 \cdot 10^{-6} \overline{R'_c} \cdot \sqrt{S_{AT}} - 4.6 \cdot \overline{R'_c}{}^2] \qquad (7.12)$$

$$V_{CB} = N_{CT}[2.39 \cdot 10^{-4} \overline{R'_c}{}^2 \cdot V_L^{1/3} - 1.54 \overline{R'_c}{}^3] \qquad (7.13)$$

which are derived from equations (7.1), (7.2), (7.10) and (7.11). The numerical coefficients in these equations depend on the geometric shape of the capillary segments, as already mentioned, and on the observation that the average length of the capillary segment L_c was found to be a function of the extent of the alveolar membrane area S_{AM} [equation (7.9) and Fig. 60a]. The total number of capillary segments N_{CT} was essentially constant for all lungs, with an average value of $277 \cdot 10^9$. Equations (7.12) and (7.13), therefore, express the alveolar capillary surface and volume as well defined functions of the size of the lung (S_{AT} or V_L respectively) and the degree of capillary distention ($\overline{R'_c}$).

In Figs. 62 and 63 the computed values of S_{CB} or V_{CB}, as listed in Table VII.2, are plotted as dots. The short bars indicate the surface or volume as it was calculat-

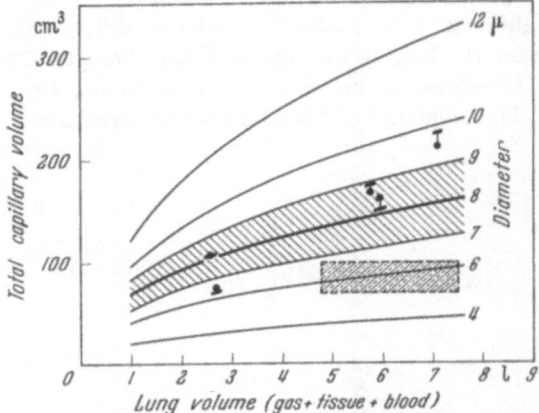

ed for each case from the generalized equations (7.12) and (7.13) on the basis of the mean capillary diameter (Table VII.1) and the size of the lung. We note satisfactory agreement between the theoretic and the computed values of S_{CB} and V_{CB}.

If, pending more precise information, we accept the values of 7 to 9 μ as the normal range of variation of the average capillary diameter the anatomic capillary blood volume of the adult lung would vary between of 100 and 200 ml, depending on the size of the organ and on the degree of vascular filling. In the normal

Fig. 63. Total internal capillary volume as function of lung volume and average capillary diameter. Symbols as in Fig. 62. Shaded rectangular area indicates range in which capillary blood volume was estimated by physiologic methods (compare text) [From WEIBEL, 1962]

living organism this anatomic capillary volume might be expected to correspond to the capillary blood volume during heavy exercise.

D. Relation of Capillary Network Units to Pre- and Postcapillaries

The capillary network was described as a large vascular sheet about 35 to 40 m² in extent, which is more or less continuous through a major part of the lung (cf. p. 75). Numerous precapillaries and postcapillaries, which represent the entrance and exit points of the capillary bed lead into this sheet. Two typical examples of such connections of the capillary network to the vessels of the transitory and conductive zones (Figs. 4 and 6; p. 8) are shown in Figs. 49 and 50. The transitory vessel is seen either to branch terminally in the form of a fan (Fig. 49), or to give off numerous lateral connectors to the capillary network, as in Fig. 50. It is not yet known whether these two patterns of branching are typical of the two types of vessels, pre- and postcapillary, or whether they merely represent variants of either type.

From our special point of view we would be interested in knowing the number of pre- and postcapillaries in the entire lung, since this would give us an indication of the distance from inflow to outflow point of the capillary network. Some preliminary data indicate that the number of precapillaries is roughly 200 to 300 million. On this basis, a capillary network unit would cover an area of inter-alveolar septa of about $15 \cdot 10^{-4}$ cm² and comprise some 1,000 capillary segments. If we assume that it covers a somewhat circular disk-like area, the diameter of this unit would be approximately 300 to 500 μ. On tracings of whole networks, as shown in Fig. 52, it is possible to follow the capillaries over distances of a few hundred micra, starting from a precapillary, without meeting another pre- or postcapillary. Therefore, the size of the capillary unit, as assumed, appears reasonable. Before conclusive statements can be made, however, this difficult problem needs to be more thoroughly investigated.

E. Discussion of the Findings on the Total Capillary Network

Various authors have described the alveolar capillary network as built up of rather dense and fine meshes (MILLER, 1937; LOOSLI, 1937; JUNGHANSS, 1958; PUMP, 1962; and others). However, we are not cognizant of any study which attempted to estimate its extent by morphologic methods. MÜLLER (1945) made some deductions on volume and surface of the pulmonary capillary bed in man, based on rather scanty information then available to him. He assumed the capillary to be a long, cylindric tube of 10 μ diameter with only a fraction of its surface being exposed to air. For an estimated capillary volume of 185 cm³ he found a respiratory surface area of 30 to 50 m².

Led by ROUGHTON, physiologists have attempted to estimate the pulmonary capillary blood volume by determining various components of the "pulmonary diffusing capacity". By means of these methods ROUGHTON (1945), MCNEILL et al. (1958), and FORSTER (1959) found a maximal capillary volume of 95 cm³ in normal individuals, which could increase to 150 to 280 cm³ in persons with pulmonary vascular congestion (MCNEILL et al., 1958). LEWIS et al. (1958) found an average normal volume of only 65 cm³. Our morphometric studies indicate that these estimates are too low to represent the anatomic capillary volume. It may, however, be possible that the instantaneous capillary blood volume is smaller than the anatomic capillary "capacity" actually determined in this study. The physiologic measurements necessary for estimating the capillary blood volume by the the method of ROUGHTON and FORSTER are obtained during a single deep breath which is held for about 10 seconds. Wearn et al. (1926, 1934) had indicated that pulmonary capillary blood flow may be intermittent so that, at a given moment, only part of the network may be filled with blood in the functioning lung. This may also be suggested by more recent work of STAUB (1961) and COSENTINO et al. (1961) on the cat lung. In addition, the depth of the breath during which the examination is carried out may tend to reduce the capillary blood volume as suggested by the studies of MACKLIN (1948), ALTMANN (1954), RILEY (1958), and PROCTOR (1961). Recently, however, BATES et al. (1960) have shown that, using a physiologic method similar to that of ROUGHTON and others, the volume of the normal capillary bed can be estimated at 100 to 200 ml, with an average of about 153 ml. These figures closely compare with the "capillary capacity" determined in this study. Nevertheless, this point needs further investigation before a final conclusion can be reached.

Errors which may have entered the present investigation should also be considered. The calculations were necessarily based on some geometric simplification of the network structure. As discussed on p. 77, this may have resulted in a degree of underestimation of the capillary volume since the "smoothed" corner regions were neglected; on the other hand, it is possible that we overestimated the volume of extremely short segments which probably have an elliptic cross-section (cf. p. 78). It is likely that these two errors largely cancelled each other out; any small residual error would probably tend to be in the direction of an underestimate.

Another error may have arisen in the fact that the capillary network was not really randomly sampled. For technical reasons only double-faced interalveolar facets could be used for this investigation. Those parts of the capillary network

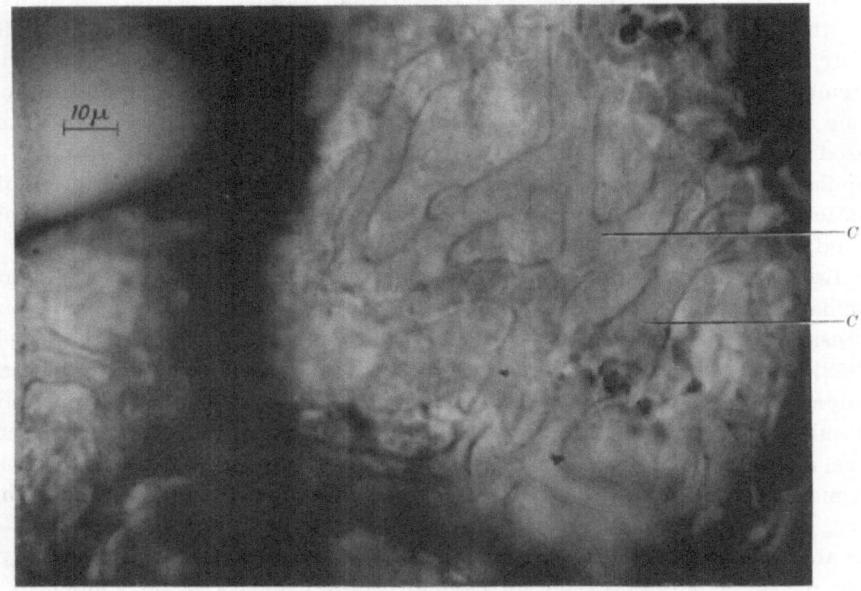

Fig. 64. Demonstration of capillary network in alveolar wall adjacent to peribronchial connective tissue by PAS reaction in basement membrane. Note that meshes are considerably wider than in double-faced interalveolar septa (Figs. 49 and 50). C = capillary. × 800

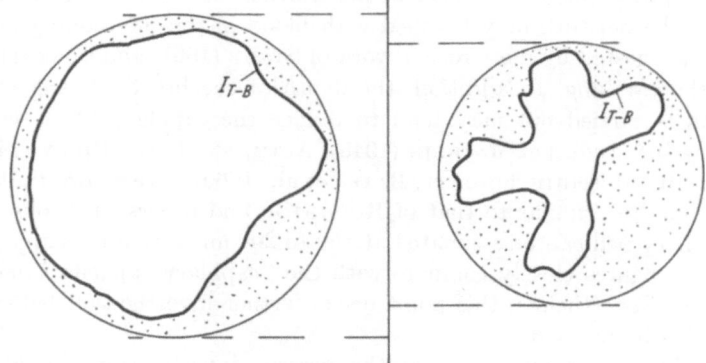

Fig. 65. Effect of different degrees of vascular filling on the extent of the tissue-blood interface I_{T-B}

which were in contact on one side with connective tissue structures (for example, the peribronchium or larger vessels), could not be assessed by our methods because the dense adjacent structures interfered with the outline of the capillaries. However, it is known (MILLER, 1937) that the capillary network of such single-faced alveolar walls has coarser meshes than is true of the double-faced interalveolar septa (Fig. 64). We can roughly estimate that such walls contain, on the unit area, about half as much capillary blood as double-faced septa. But these alveolar walls also contributed only the area of one face to the estimate of the alveolar surface, while interalveolar septa contributed two sides. By using $S_{AM} = 1/2 \cdot S_{AT}$ as the basis for estimating the extent of the capillary network (cf. p. 79), we compensated, to a great degree, for any error which would have been introduced by disregarding these marginal areas. In addition, the capillary network of these regions represents only a small fraction of the entire capillary bed. On the whole, we would estimate possible residual errors from the two sources mentioned as being $\pm 10\%$ or less.

Geometric simplifications may have caused the capillary surface area S_{CT} to be a slight underestimate of the tissue-blood interface I_{T-B} due to the surface texture of the capillary endothelium (Fig. 65). This underestimate is probably not as large as that involving the air-tissue interface. It seems appropriate to let

$$I_{T-B} = \delta \cdot S_{CT} \approx 1.1 \cdot S_{CT} . \tag{7.14}$$

The overall capillary dimensions will also be influenced by variations in the functional state of the lung. Varying degrees of vascular injection have been shown to influence both volume and surface of the capillary bed. We may, how-ever, anticipate that the tissue-blood interface I_{T-B} may be but slightly in-fluenced by this variation; thus, we would have to assume that the coefficient δ is an inverse function of the degree of vascular injection. This assumption appears reasonable, as shown in Fig. 65.

As already mentioned, various studies suggest that different degrees of infla-tion tend to influence the overall capillary dimensions. Really deep inflation appears to empty a portion of the capillaries (ALTMANN, 1954; RILEY, 1959; PROCTOR, 1961). This point calls for further quantitative investigation.

Chapter VIII

Composition and Dimensions of Alveolo-capillary Tissue Barrier

A. Nature of the Alveolo-capillary Tissue Barrier

The alveolar capillary blood is separated from the alveolar air by a layer of tissue (Fig. 66) which represents a resistance to the diffusion of gases between air and blood. This so-called air-blood barrier is composed of an alveolar epithelium and a capillary endothelium, separated by an interstitial framework of basement membranes and connective tissue elements (Fig. 67). Over large areas this barrier is greatly attenuated so as to form a thin alveolo-capillary "membrane", while in other regions it may develop considerable thickness (Figs. 66 and 67). Various electron microscopic studies have revealed that each of the three constituent layers[1] is continuous throughout the lung (Low, 1953; KARRER, 1956; BARGMANN and KNOOP, 1956; SCHULZ, 1957; POLICARD et al., 1957; and others). The fine structure of their components has been discussed in detail. In the following pages these findings will be briefly reviewed in order to define the morphologic entities we have analyzed quantitatively. The illustrations used for this purpose (Figs. 66

---

[1] Of the interstitial layer only the basement membranes are truly continuous, as will be shown.

to 70) are taken from rat lungs because better tissue preservation can be achieved in material from experimental animals than in human lungs. Figure 71 shows a

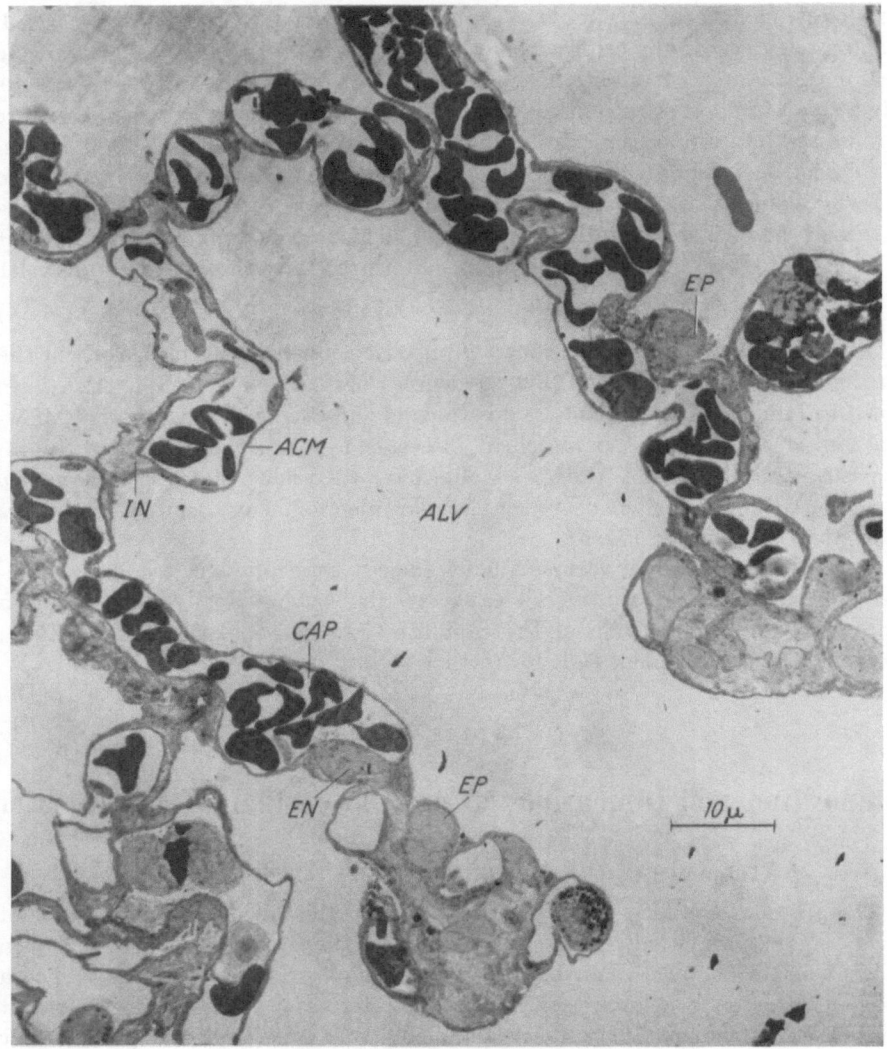

Fig. 66. Section of alveolus of rat lung, showing relation between alveolar air and capillary blood. The alveolo-capillary tissue layer is very thin ("membranous") over large areas (*ACM*), but is thicker where it contains alveolar epithelial (*EP*) and capillary endothelial cells (*EN*), or interstitial elements (*IN*). × 1350

section of a human alveolar capillary for comparison. We shall show that species differences are not qualitative but quantitative. We can, therefore, generalize in the description of the components of the air-blood barrier.

1. Alveolar Epithelium

The alveolar epithelium is composed of three cell types. The nucleus of *small epithelial cells* (POLICARD et al., 1959) is enveloped by a thin shell of cytoplasm which contains a few mitochondria and a moderate amount of endoplasmic reticulum (Figs. 66 + 35). It rests upon the epithelial basement membrane and is

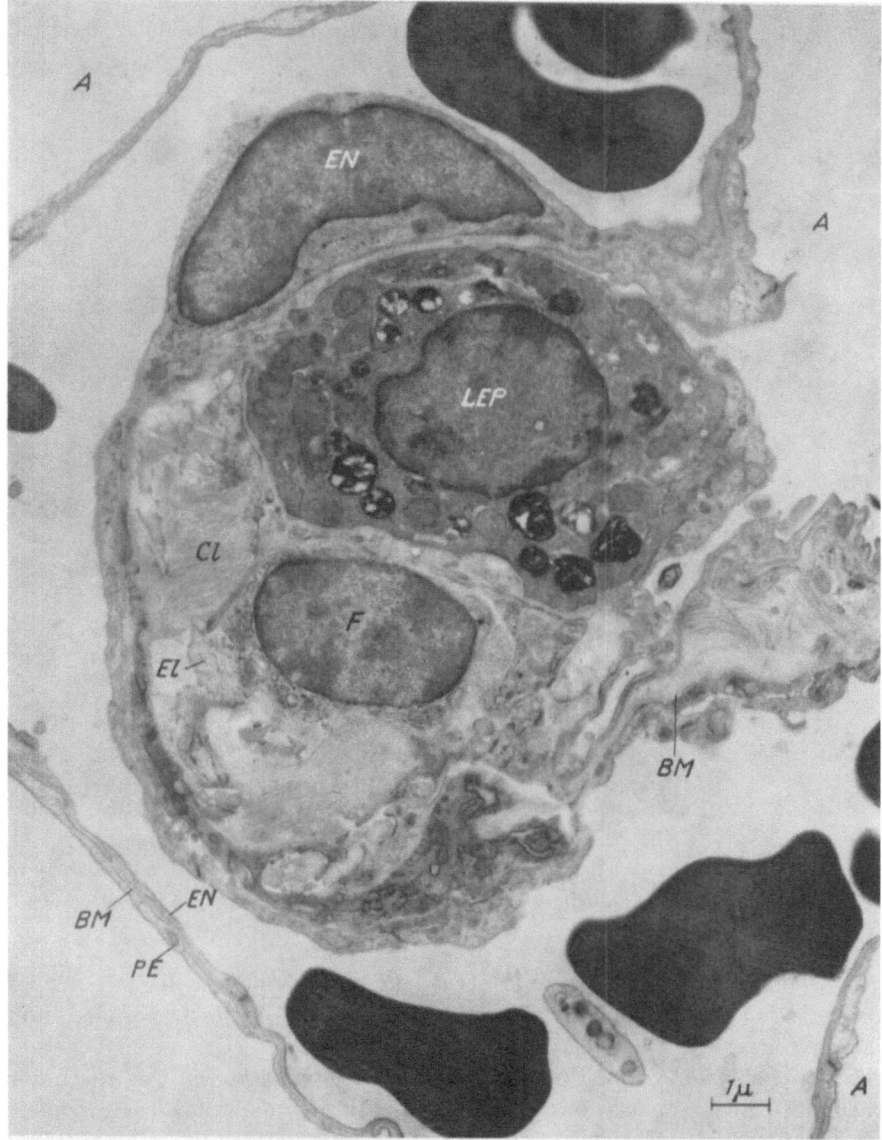

Fig. 67. Oblique section of mesh of capillary network of rat lung showing major constituents of alveolo-capillary tissue space, which are most frequently found in center of mesh. Nucleated portions of endothelial cell (*EN*), of large epithelial cell (*LEP*), and of fibroblast (*F*) are shown, as well as collagenous (*Cl*) and elastic (*El*) fibers. Basement membrane (*BM*) is recognized beneath endothelium and epithelium in thick and thin portions. A = alveolus, capillary marked by erythrocytes. × 7700

found mostly in the central depression of capillary meshes (Fig. 66). At its basal circumference the cytoplasm attenuates (Fig. 35) and extends laterally over large areas as a thin cytoplasmic layer (0.2 to 0.5 μ thick) which contains small membrane-bounded vesicles (Figs. 69 and 70), some endoplasmic reticulum and occasional mitochondria. The internal and external plasma membranes show numerous small and rounded invaginations, suggesting pinocytotic activity (Fig. 70). The margins of adjacent epithelial cells form intercellular junctions

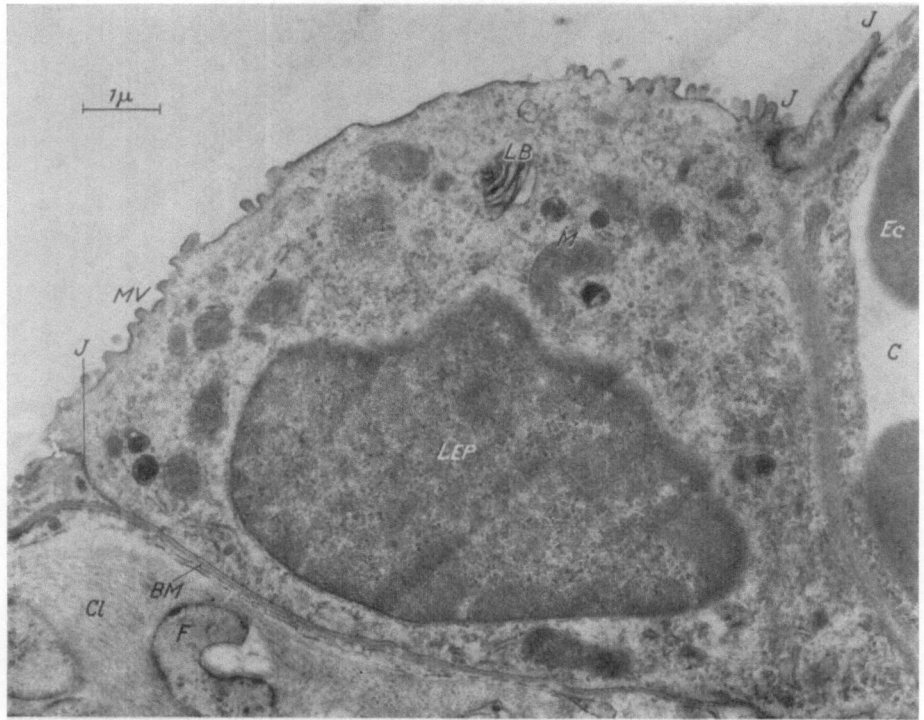

Fig. 68a. Paramedian section of large alveolar epithelial cell (*LEP*) which rests on basement membrane (*BM*) in central depression of capillary mesh (*C*) and contains typical lamellated body (*LB*), as well as mitochondria (*M*) Free surface shows short microvilli (*MV*). At the periphery the cell is connected to attenuated processes of adjacent small epithelial cells (*J*). Underlying interstitium contains bundles of collagen fibrils (*Cl*) and processes of fibroblast (*F*). Rat lung. × 10,000

(Figs. 68 and 69) similar to those observed in vascular endothelia (MOORE and RUSKA, 1957; and others); the cells thus form an uninterrupted epithelial lining of the alveolus.

The *large alveolar cells* of POLICARD (1959) are interspersed in this continuous epithelium. They are characterized by a more abundant perinuclear cytoplasm (Figs. 67 and 68a), relatively rich in endoplasmic reticulum and containing numerous mitochondria, along with lamellated bodies repeatedly described under various names (SCHLIPKÖTER, 1954; SCHULZ, 1959; POLICARD et al., 1959; and others) and which appear to be specific for these cells. The large cells do not form attenuated lateral extensions; the processes of surrounding small alveolar cells are in close contact with the body of the large cells through a regular intercellular junction (Fig. 68a). The free surface of these cells carries numerous short microvilli.

The *free alveolar cells* or alveolar macrophages are believed to originate from the large alveolar cells (POLICARD, 1959). They are not in contact with the basement membrane but are superimposed on the uninterrupted alveolar epithelium, with the surface of which they are in close but loose contact (Fig. 68b). They extend cytoplasmic processes of varying length which show signs of phagocytosis (KARRER, 1958 and 1960; POLICARD, 1959; and others). The perinuclear cytoplasm

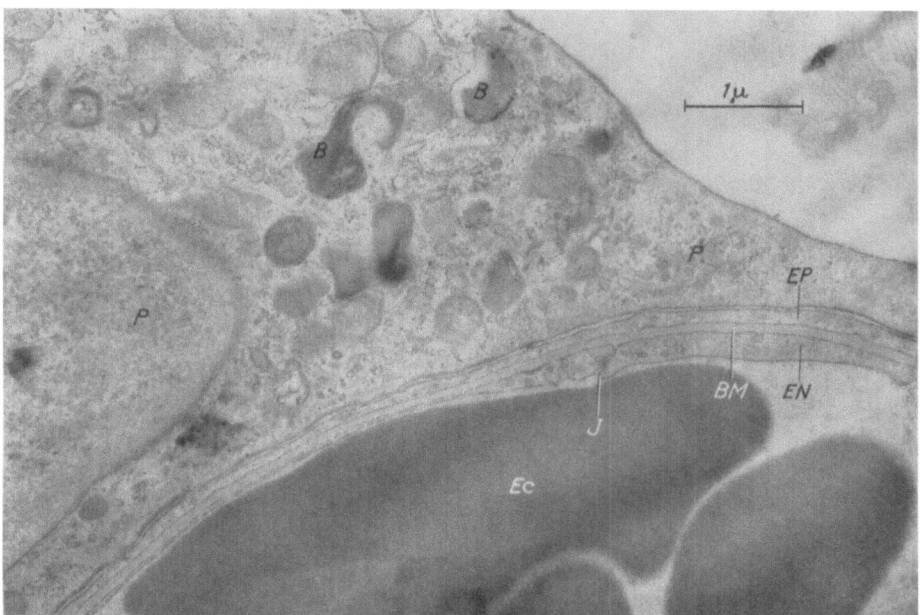

Fig. 68b. Alveolar macrophage (*P*) contains varied inclusion bodies (B), and lies on attenuated portion of epithelial cell (*EP*), which builds up actual air-blood barrier together with endothelium (*EN*) and interstitium (*BM*). *J* = intercellular junction. Capillary contains erythrocytes (*Ec*). Rat lung. ×16,000

is rich in endoplasmic reticulum and mitochondria, and contains inclusion bodies of varied nature (KARRER, 1958) which appear to be the result of phagocytosis.

2. Capillary Endothelium

The endothelium forms a poreless, uninterrupted lining of the alveolar capillaries (KARRER, 1956b; BARGMANN and KNOOP, 1956; POLICARD et al., 1957; and others). The relatively sparse perinuclear cytoplasm contains a few mitochondria, some elements of the Golgi complex, and a moderate amount of endoplasmic reticulum with ribosomes (Fig. 69a). Peripherally the cytoplasm becomes attenuated and extends as a thin veil (0.2 to 0.5 μ thick) over large areas, always following the endothelial basement membrane. The plasma membranes show small invaginations about 700 Å in diameter, and the cytoplasm contains vesicles of the same size. It is assumed that these formations are related to pinocytosis (DE GROODT et al., 1958). Occasional mitochondria may be observed in this attenuated portion of the endothelium, predominantly towards the periphery. Adjacent cells are connected in the form of simple intercellular junctions whereby the two cells

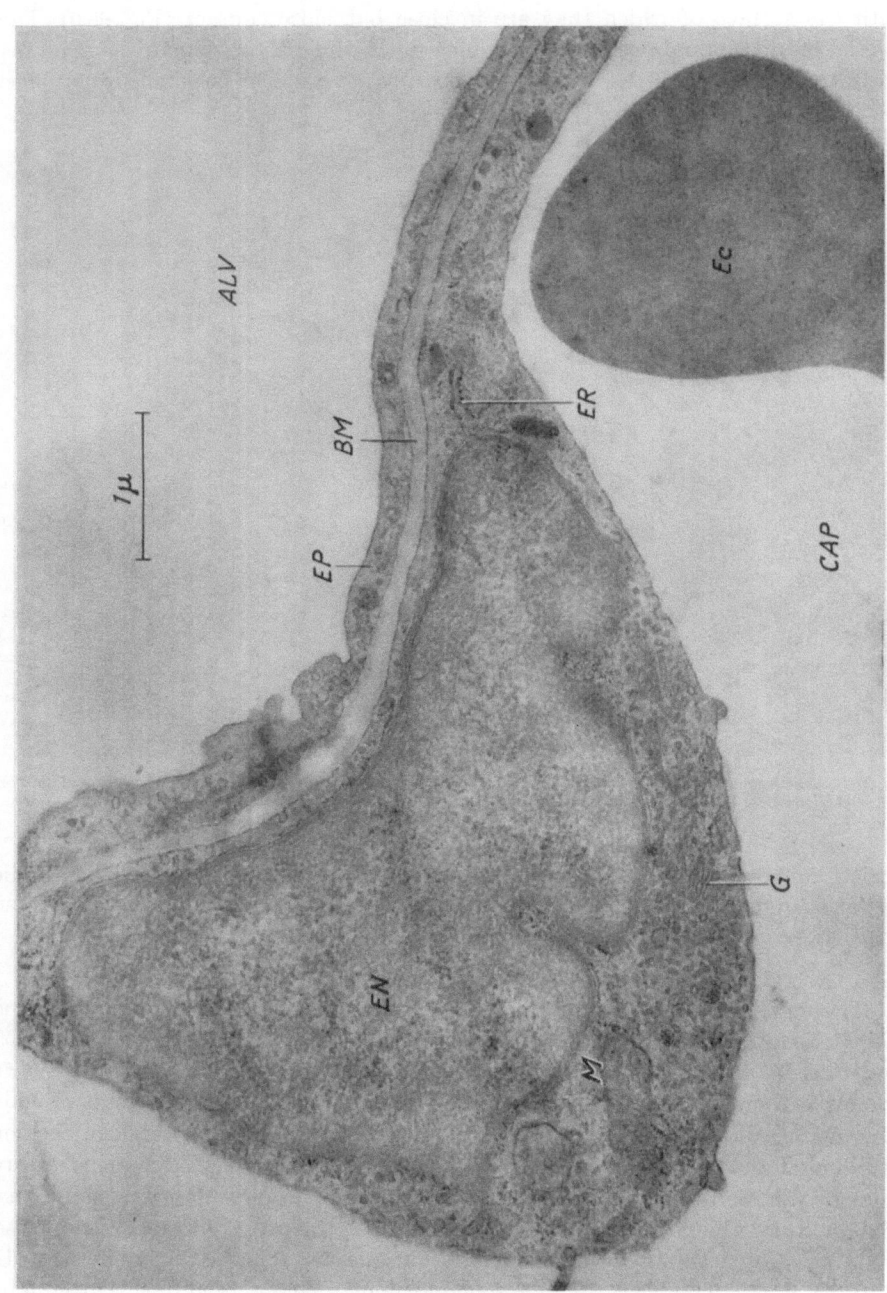

Fig. 69a. Alveolo-capillary tissue layer of rat lung containing nucleated portion of endothelial cell (*EN*). Perinuclear cytoplasm contains mitochondria (*M*), endoplasmic reticulum (*ER*), and elements of the Golgi complex (*G*). Cytoplasmic veil of small epithelial cell (*EP*) covers narrow interstitial space (*BM*). *Ec* = erythrocyte in capillary lumen. × 20,000

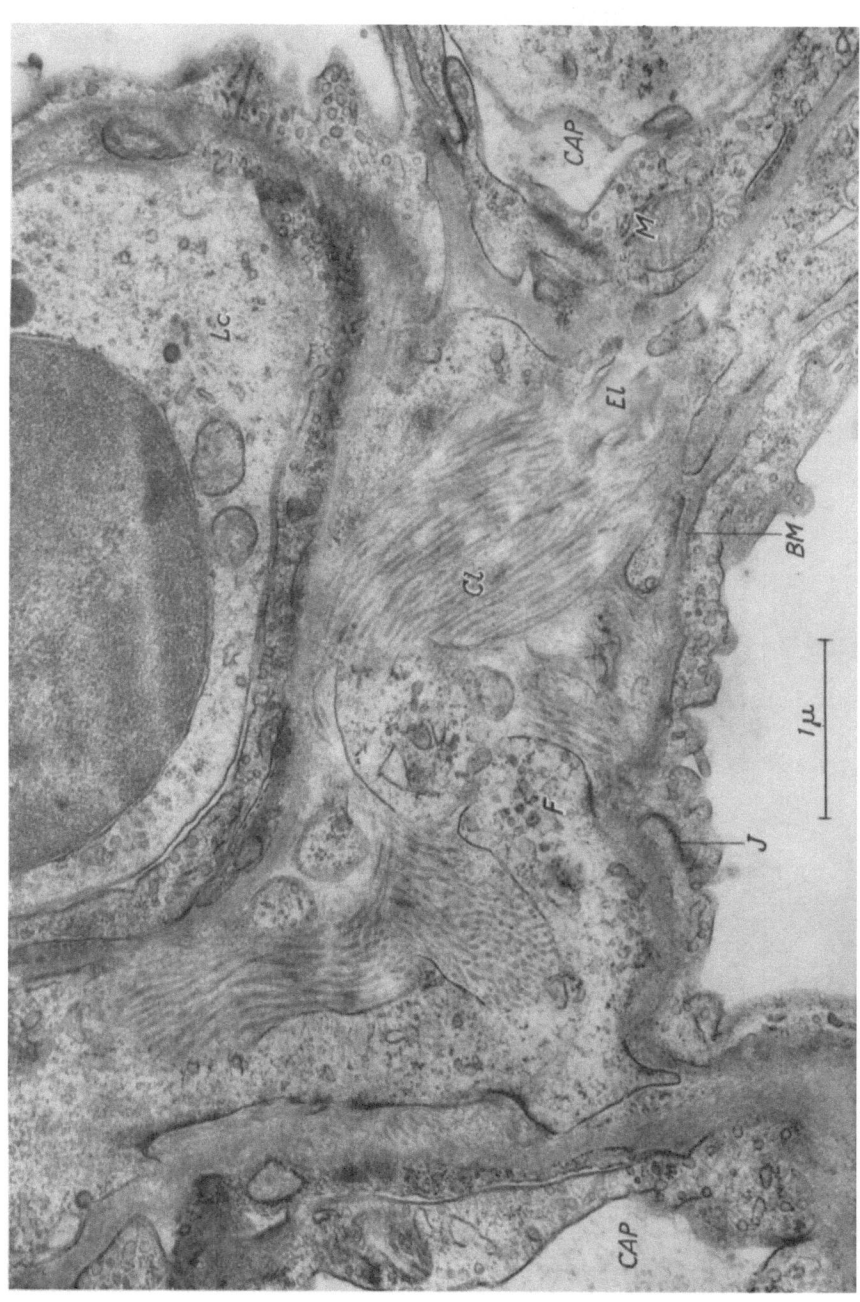

Fig. 69b. Wider interstitial space between three capillary segments (*CAP*) contains elastic fibers (*El*) and bundles of collagen fibrils (*Cl*) which appear closely related to processes of a fibroblast (*F*). Basement membranes (*BM*) are seen beneath endothelium and epithelium. Attenuated portion of endothelium contains a mitochondrion (*M*) near intercellular junction (*J*). One capillary contains a lymphocyte (*Lc*). × 24,000

frequently overlap slightly (Figs. 68b, 69b, 70a, and 71). No special structures such as "terminal bars" can be observed in association with these junctions.

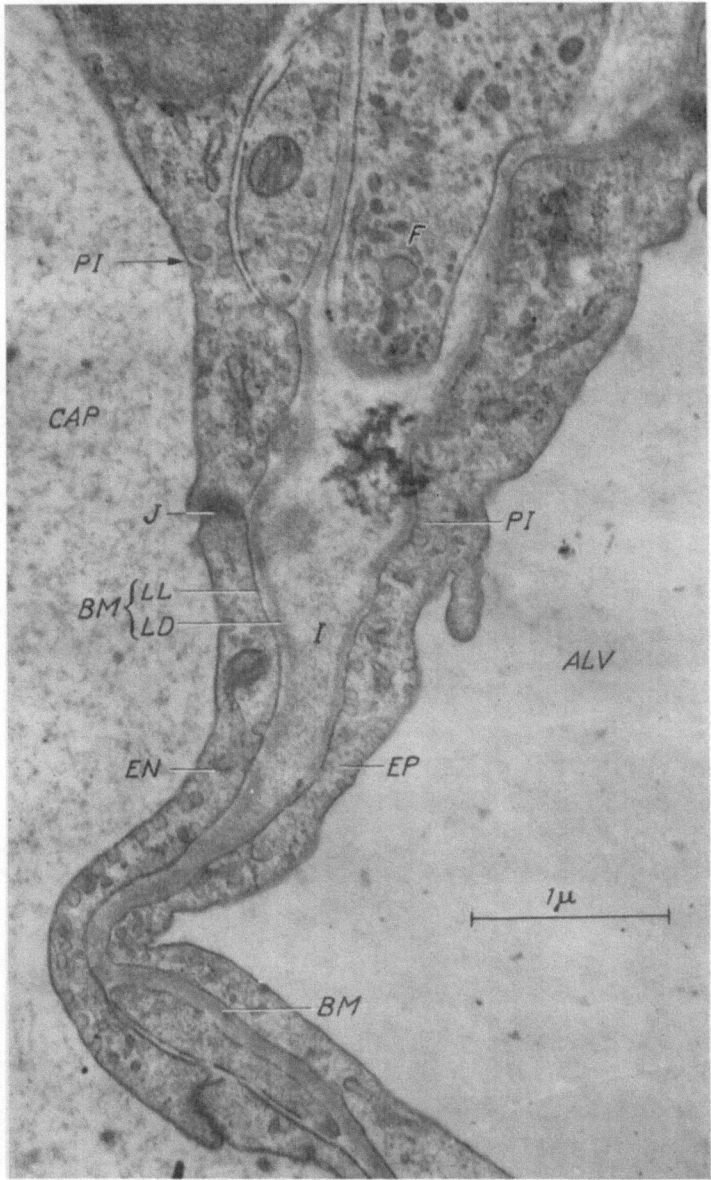

Fig. 70a. Transition between narrow and wider interstitial space (*I*) of alveolo-capillary tissue barrier of rat lung. Basement membranes (*BM*) follow epithelium (*EP*) and endothelium (*EN*) and show light (*LL*) and dense (*LD*) layers. Epithelium and endothelium show signs of pinocytosis (*PI*). Interstitial space contains fibroblast (*F*).
× 26,000

3. Interstitium

The interstitial space is limited on one side by the endothelial, on the other side by the epithelial, basement membrane. Each basement membrane is separated from the endothelial or epithelial cells, respectively, by a light "laminar" space (Fig. 70a). In the thinnest portions of the air-blood barrier the epithelial and

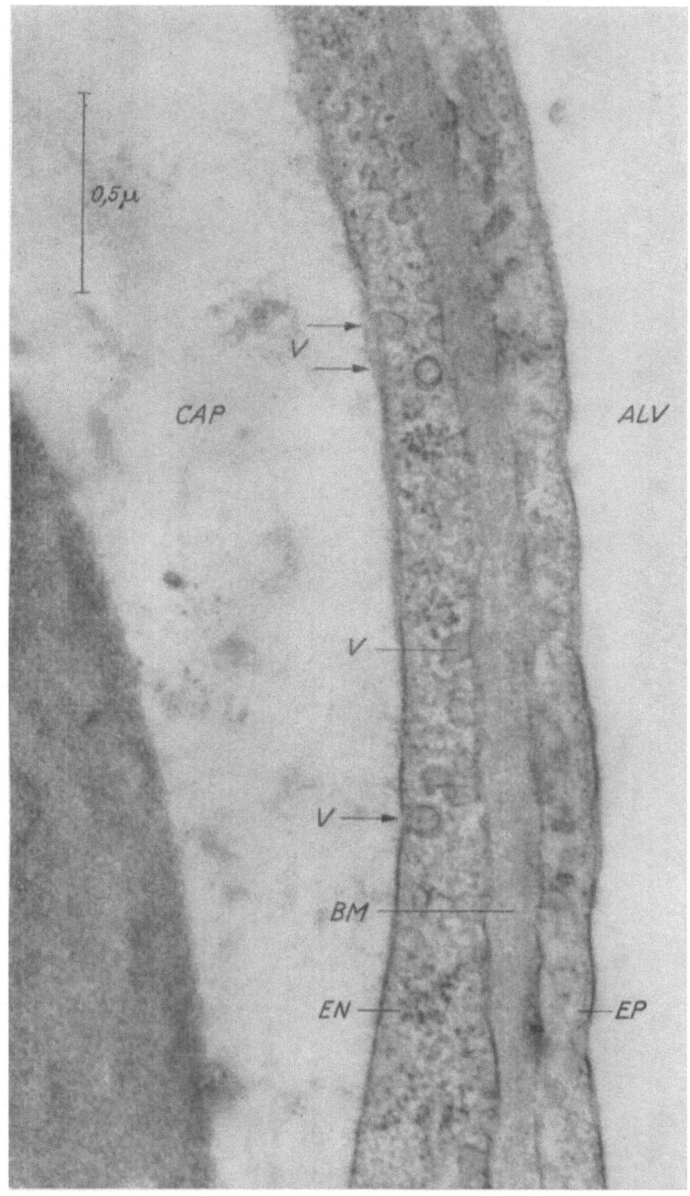

Fig. 70b. Thin "membranous" portion of air-blood barrier of rat lung showing the continuous layers of epithelium (*EP*), endothelium (*EN*), and interstitial basement membranes (*BM*). Endothelium shows pinocytotic vesicles in connection with internal and external plasma membranes (*V*). Capillary contains erythrocyte. × 55,000

endothelial basement membranes are so closely approximated that they become indistinguishable (Fig. 70); in thicker parts they follow the epithelial or endothelial cells (Figs. 69b and 70a). The space thus bounded by the basement

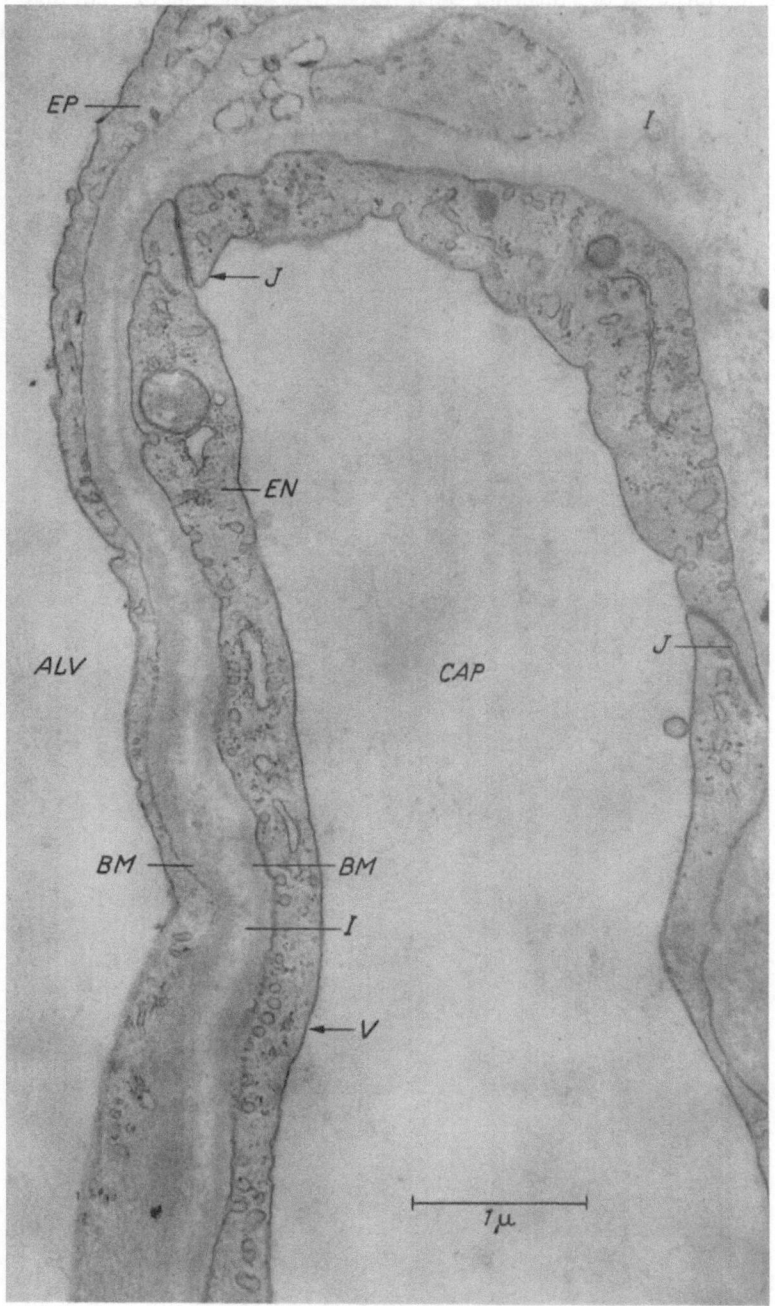

Fig. 71. Alveolar capillary of human lung (*CAP*). Air-blood barrier contains essentially the same components as that of the rat. Small alveolar epithelial (*EP*) and endothelial cells (*EN*) form attenuated cytoplasmic veils which compose "membranous" portion of barrier; the interstitial space (*I*) shows the two basement membranes (*BM*), which are separated in this instance. *ALV* = alveolus, *V* = pinocytotic vesicles, *J* = intercellular junctions.
× 23,000

membranes contains, within a loose "ground substance", bundles of collagenous fibers, elastic fibers, some fine microfibrils (Low, 1961), and fibroblasts which appear to extend with long processes along the fibrous elements (Figs. 67, 68a, and 69b). Occasionally the interstitium also contains some free cells, for example, interstitial macrophages or plasma cells. All interstitial cells are free, i. e., none is covered by basement membrane material.

4. Discussion

This rather complex structure of the alveolo-capillary tissue layer can be summarized by attributing its constituents to one of the following two classes which will influence diffusion in a somewhat different manner (Fig. 72). This

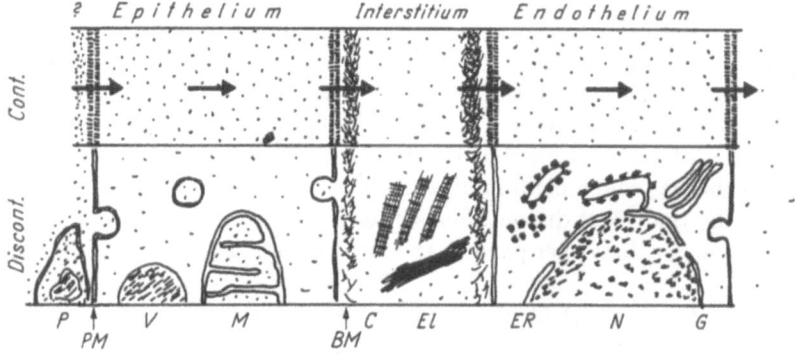

Fig. 72. Schematic representation of alveolo-capillary tissue space (air-blood barrier) which is subdivided into continuous and discontinuous components. Plasma membrane (*PM*) and basement membrane (*BM*) are continuous. Alveolar macrophages (*P*), vesicles and vacuoles (*V*), mitochondria (*P*), endoplasmic reticulum (*ER*), nuclei (*N*) and Golgi complex (*G*) are discontinuous components, as well as collagenous (*C*) and elastic (*El*) connective tissue fibers

subdivision may, to some extent, be artificial since even the "continuous" components are "discontinuous" at the molecular level. The classes differ, therefore, by the degree of coherence of the units.

a) Continuous or Coherent Layers

These layers are necessarily crossed by gas molecules moving from the air into the blood and vice versa, since they are ubiquitary throughout the gas exchange region. This minimal air-blood barrier is composed of two cellular layers separated by an interstitial space. Molecules in transit must thus cross four cell membranes and two layers of *cytoplasmic matrix* which is supposedly a sol of proteins, carbohydrates, salts, and so forth. The *plasma membranes* of epithelium and endothelium are about 70 Å thick and display the characteristic structure of a double leaflet of phospholipids (ROBERTSON, 1960). The *basement membranes* reveal a dense, finely granular structure with some microfibrillar components (Low, 1961). The composition of the *interstitial ground substance* probably corresponds to that of the extracellular fluids in general but may contain some mucopolysaccharides in addition. An additional continuous coat of the air-blood barrier may be formed by a possible fluid layer covering the surface of the alveolar epithelium (CHASE, 1959), although the nature and dimensions of this film are not yet defined.

7*

b) Discontinuous Components

The other components of the alveolo-capillary tissue layer can be considered more or less as discrete "objects" which may act as obstacles to the diffusing gases, but which do not necessarily have to be traversed, since there will always be a "free" path leading around them. However, the relatively impenetrable discontinuous components will tend to reduce the useful cross section of the diffusing path.

Within the cytoplasm discrete components include dense ribonucleoprotein particles (ribosomes) which are in part free and in part attached to membranes bounding the cisternae of the endoplasmic reticulum. Small pinocytotic vesicles and cisternae of endoplasmic reticulum and Golgi complex may not necessarily offer serious resistance to diffusion except for their lipoprotein membranes. Mitochondria, nuclei and larger inclusion bodies form larger obstacles; their often dense content may also have a high diffusion resistance. In the region of intercellular junctions the adjacent cells overlap to some extent, so that two additional plasma membranes are intercalated in the diffusion path.

Within the interstitium we find bundles of collagen fibrils, elastic fibers and ramified fibroblasts, which contain all the above-mentioned cellular components. It is noteworthy that these discontinuous interstitial elements are usually confined to rather narrow portions over parts of the alveolo-capillary tissue space (Fig. 66).

B. Over-all Dimensions of Air-blood Barrier

The first step in the analysis of the air-blood barrier will be to ascertain its thickness τ, which is defined as the distance from the superficial plasma membrane of the alveolar epithelium to the internal plasma membrane of the endothelium. Therefore, it excludes the postulated fluid film which may coat the free surface of the alveolar epithelium. The following preliminary report on these studies, still in progress at the time of writing, is based on findings observed in two rat lungs. The results, which were derived from a small number of samples, can nonetheless be regarded as indicative of the actual dimensions of the barrier, although they serve mainly to illustrate the aims pursued in this study.

An estimate of the variation in thickness of the air-blood barrier was first obtained by systematic measurements on electron micrographs of random samples of thin lung sections ($\sim$ 900 Å), which were obtained by the sampling procedure described on p. 50. The portions of alveolar capillaries contained within the randomly sampled meshes of the screen carrying the section were photographed at about 4,000 $\times$ electronic magnification. The entire sample field was reconstructed by montage of the prints which had a final magnification of 14,000 $\times$. At equidistant points measurements of the air-blood barrier were made along the normal to the capillary basement membrane. Some 130 measurements of τ' were recorded on two samples from one rat lung. The frequency distribution of τ' is plotted in Fig. 73; the highest density was found between 0.4 and 1.0 μ, with a gradual decrease toward 2 μ and a flat tail up to 5 μ and more. This distribution pattern results partly from the variation of the apparent thickness of the barrier due to oblique sectioning, partly from an actual variation of the true thickness from point to point. Thus far, no attempt has been made to separate these two superimposed distributions. Undoubtedly, however, this will be necessary. It can be anticipated that the distribution of actual thickness will be somewhat narrower than that of the apparent ones, as indicated in Fig. 73.

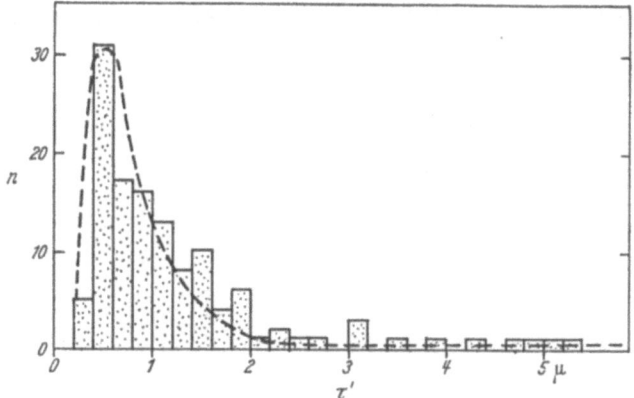

Fig. 73. Frequency distribution of thickness τ' of air-blood barrier as measured along the normal to the capillary basement membrane. Broken line indicates estimated distribution of "true" thicknesses after allowance for oblique sectioning. Compare text

The actual distribution of over-all thicknesses τ of the air-blood barrier is characterized by two parameters which have a different signification: the arithmetic mean

$$\bar{\tau} = \frac{\sum\limits_{\tau} \tau \cdot n_\tau}{\sum\limits_{\tau} n_\tau} \tag{8.1}$$

and the harmonic mean τ_h which is the inverse of

$$\frac{1}{\tau_h} = \frac{\sum\limits_{\tau} \frac{1}{\tau} n_\tau}{\sum\limits_{\tau} n_\tau}. \tag{8.2}$$

The average thickness $\bar{\tau}$ is a measure of the mass of tissue building the barrier, while τ_h is its effective over-all thickness with respect to its diffusion resistance: the flow of a gas across a tissue sheet is, at each point, inversely proportional to the local thickness. τ_h^{-1} is the average inverse thickness and is thus directly related to the average or over-all diffusion conductance of the barrier. It should be noted that $\bar{\tau} > \tau_h$, the difference being the greater the wider the spread of the distribution of τ.

On p. 30 we presented two principles by which $\bar{\tau}$ and τ_h can be determined by random measurements obtained on the fluorescent screen of an electron microscope. While the practical applicability of these principles is still being tested in model experiments, $\bar{\tau}$ and τ_h have been estimated in a preliminary fashion on one rat lung. The random samples used for this study were again obtained by the procedure outlined on p. 50.

The arithmetic mean of the thickness $\bar{\tau}$ was computed by equation (3.70) from the number p of end points of the sampling lines (Fig. 21) of length $z = 3 \cdot 10^{-4}$ cm lying on the air-blood barrier, and the number n of intersections of these lines with the internal and external boundaries of the barrier (see p. 32). Figure 74a shows the cumulative estimates of the average of the ratio p/n as obtained on 20 sample fields, whereby a total of 60 sampling lines were placed on each field. The small deviations of the estimates of the average of cumulated successive samples from

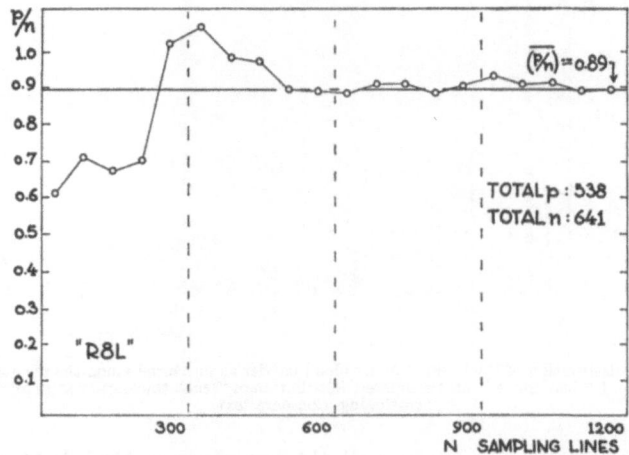

Fig. 74a. Determination of $\bar{\tau}$. Average of ratio p/n obtained by cumulation of successive samples. Compare text

the final average reveal that the result is satisfactorily reliable for our present purpose. The half-width of the 95% confidence interval is about 10% of the mean. On the basis of these data we derive that $\bar{\tau} = 1.30\ \mu$.

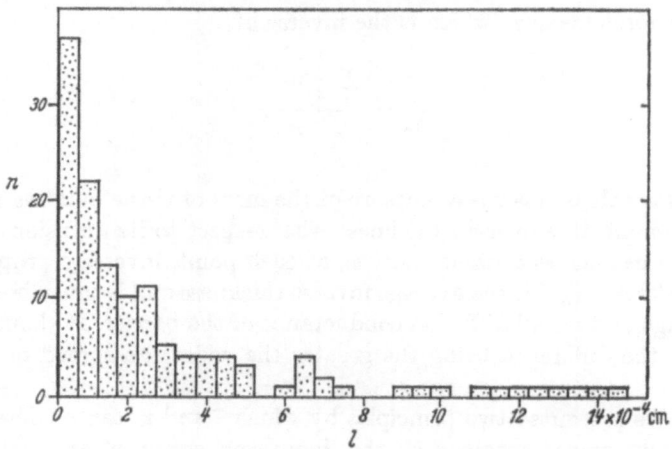

Fig. 74b. Determination of harmonic mean thickness τ_h from length l of random intercepts. Distribution of l obtained on same animal

The harmonic mean thickness of the air-blood barrier τ_h was determined on the same samples according to the principle outlined on p. 33. A row of equidistant dots (cf. Figs. 19 and 21) was randomly superimposed on the sections. The frequency distribution of the lengths l of these lines crossing the air-blood barrier is given in Fig. 74b. From these data τ_h was computed by equation (3.80), and found to be $\tau_h = 0.46\ \mu$. Model experiments have revealed, however, that the curvature of the air-blood barrier may have caused a systematic underestimate of τ_h by 10 to 20%. It appears therefore appropriate to suggest, at present, the range $0.45\ \mu < \tau_h < 0.55\ \mu$, in which the actual harmonic mean thickness of the air-blood barrier may lie.

C. Distribution of Components Within Alveolo-Capillary Tissue Space

The major components of the alveolo-capillary tissue space, which were described as being either continuous or discontinuous, have been analyzed quantitatively by two different approaches. The method of point counting on the fluorescent screen of the electron microscope (see p. 21) yielded the volumetric fractions of the entire tissue space occupied by alveolar epithelium, interstitium and capillary endothelium. The results obtained on one rat lung on a total of 1,114 points are listed in Table VIII.1. The interstitium occupies about 40% of the volume, while each of the two continuous cellular layers amounts to about 30%.

Table VIII.1. *Volumetric Composition of Alveolo-Capillary Tissue Space*

Epithelium	0.32
Interstitium	0.40
Endothelium	0.28

Total number of points counted: 1114

Table VIII.2. *Volumetric Composition of Air-Blood Barrier Measured as Fractional Length of Normal Path*

Epithelium	0.37
Interstitium	0.21
Endothelium	0.42

Total number of sample lines: 75

Another estimate of the volumetric composition of the air-blood barrier was arrived at by fractional measurements along the equidistant normals drawn on electron micrographs of randomly sampled sections of a rat lung (see above). The results are listed in Table VIII.2. By this method the interstitium is found to occupy only about 20% of the tissue space, as compared with 40% found by point counting. This may be due partly to the fact that the systematic measurements on normals of the capillary basement membrane will tend to underestimate the contribution of the septa filling the intercapillary meshes. These regions, however, show a relatively large amount of connective tissue (Figs. 66 and 67).

D. Discussion

Various electron microscopists have estimated some dimensions of the alveolo-capillary tissue layer, such as the approximate thickness of its thinnest and thickest regions (Low, 1953; BARGMANN and KNOOP, 1956; KARRER, 1956; POLICARD et al., 1957; SCHULZ, 1959; MEESSEN, 1960; and others). MEESSEN (1960) has also indicated the fractional depth of epithelium, interstitium and endothelium in the thinnest regions of the air-blood barrier in various species.

All these studies, however, have drawn an implied or clearly stated distinction between an "effective" (thin) and an "ineffective" (thick) portion of the air-blood barrier, without indicating the proportion between these two ill-defined parts. In the perspective of this study it appears essential, therefore, to regard the thickness of the air-blood barrier as continuously varying between its minimum and maximum values. Figure 75 shows the distribution of barrier thicknesses τ on the unit area of capillary surface of a rat lung as it could be derived from the measurement of normal paths described (Fig. 73). It demonstrates that throughout about half the area τ is smaller than its over-all harmonic mean value, while in the remaining half it is gradually increasing.

Analysis of fractional lengths of the normal path revealed that the continuous components were present in the same proportion in thin and thick areas (Fig. 75).

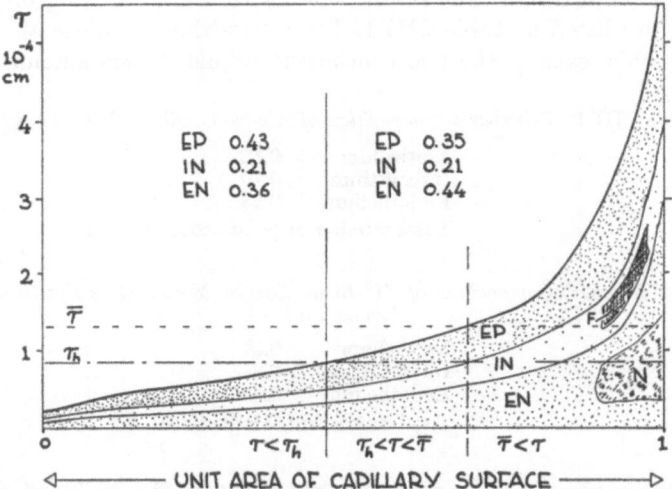

Fig. 75. Distribution of over-all thickness τ of air-blood barrier on the unit area of capillary surface, and fractional depth of the three continuous layers of epithelium (EP), interstitium (IN) and endothelium (EN). Approximate distributions of major discontinuous components (N = nuclei, F = fibers) are indicated in thick portion

It was also observed that major discontinuous components, such as connective tissue fibers or nuclei, occupied about 6 to 8% of the unit area; they were, by their nature, confined to the thicker regions. Figure 75 can thus be regarded as a tentative model of the air-blood barrier of the rat lung. This will need further refinement by study of the distribution of smaller discontinuous components, which amount to a considerable fraction of the tissue volume and are also present in the thin regions.

As stated previously, these pilot studies were carried out on rat lungs. It was shown earlier that the alveolo-capillary tissue layer of the human lung consists essentially of the same components. There is some suggestive evidence that the over-all dimensions of the human barrier may differ from those in the rat, mainly in the sense that the thicker portions occupy a larger fraction of the unit area of capillary surface. Further substantiation is necessary.

Chapter IX

Geometric and Dimensional Model of the Respiratory Zone

Physiologists repeateadly have made use of simple anatomic models of the respiratory elements of the lung in their functional analysis of gas exchange phenomena occurring at the alveolo-capillary level (MÜLLER, 1945; GRÜN et al., 1956; KREUZER, 1953; THEWS, 1957; and others). While these models were mostly hypothetic and did not reflect the actual geometric and quantitative relations between the elements, it is now possible to propose a more "realistic" model on the basis of the morphometric studies presented in the preceding paragraphs.

The alveolus is regarded as the unit of the model of the respiratory zone. The normal human lung — age 8 years or older — contains a total number $N_{AT} = 300 \cdot 10^6$ such units which have an individual volume

$$v_m = \bar{v}_A = \frac{\varrho_A \cdot V_L}{N_{AT}} = \frac{\varrho_A \cdot V_L}{3} \cdot 10^{-8} \,. \qquad (9.1)$$

These alveoli are packed around the cylindric surface of some 14 million alveolar ducts and sacs (see p. 131). They have a polyhedral shape whose various charac-

teristic features have been described on p. 56. In Fig. 41 three different typical simplifications of the alveolar shape are proposed. In designing a model of the respiratory zone mainly two of them will be useful: If single alveoli are to be considered, the truncated spheric shell of Fig. 41e and f appears most convenient because of its geometric simplicity; it will, therefore, be employed in this chapter. In later considerations on the aggregate of alveoli in relation to alveolar ducts we shall assume alveoli to be wedges of a cylindric sleeve as shown in Fig. 41a (cf. p. 140).

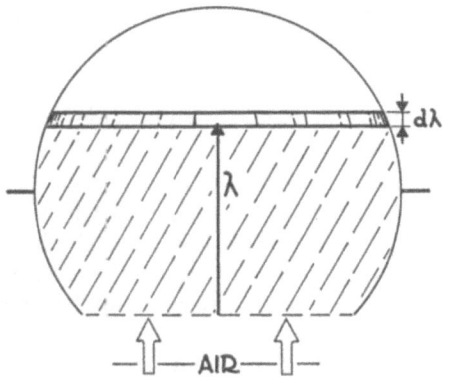

Fig. 76. Model of single alveolus. Compare text

The spheric model alveolus is shown in more detail in Fig. 76. Its depth is assumed to be 5/3 of the alveolar radius. The cut surface is open and is connected to the alveolar duct. This spheric shell has a surface-to-volume ratio $\sigma_m = 4.25$

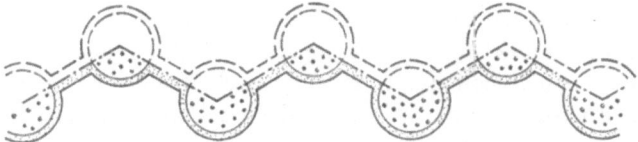

Fig. 77. Arrangement of capillaries in wall of model alveolus

[cf. equation (6.5)], which is lower than the average $\bar{\sigma}_A = 4.87$ pertaining to the alveolus (see p. 61). It will, therefore, be necessary to increase the surface of the shell by a factor of 1.15 without changing the volume in order to maintain the correct relationships; this can be achieved by surface corrugation (Fig. 77). As discussed on p. 70, the air-tissue interface is about 20% larger than the alveolar surface area. The bulging capillaries cause some additional surface corrugation which will be sufficient to provide a realistic relationship between the volume v_m of the model alveolus and the area I_m of its air-tissue interface; the latter follows from the surface s_m of the spheric shell by

$$I_m = 1.35 \cdot s_m \qquad (9.2)$$

or

$$I_m = 5.75 \cdot v_m^{2/3} . \qquad (9.3)$$

The gases enter the alveolus across the open cut surface; the depth of penetration into the alveolus can be defined as $x = \lambda r$ (Fig. 76), whereby $0 \leqq \lambda \leqq 5/3$. At a depth λ the alveolar model will have a cross section

$$A(\lambda) = \pi r^2 \left(1 - \left(\lambda - \frac{2}{3}\right)^2\right) \qquad (9.4)$$

and the volume of air from the entrance plane to this depth λ will be

$$V_m(\lambda) = \frac{\pi}{3} r^3 \left(\frac{13}{3} \lambda - 2 \lambda^2 - \lambda^3\right) . \qquad (9.5)$$

Between the depths λ and $(\lambda + d\lambda)$ a ring of the spheric shell of area

$$ds_m = 2\pi r^2 \cdot d\lambda \qquad (9.6)$$

will be in contact with air (Fig. 76). It is observed that ds_m is independent of the depth λ.

The element of the alveolar capillaries was shown on p. 76 to be a short cylindric capillary segment which forms one chord of the hexagonal meshwork illustrated in Figs. 49 to 53. The lung was found to have a total of some $280 \cdot 10^9$ such segments enclosed in the interalveolar septa. The shell of each model alveolus would, therefore, carry about 1,800 capillary segments, so that only their interior half will participate in the gas exchange with the particular alveolus, while the opposite half will be in contact with adjacent alveoli (Fig. 77).

On the model alveolus the capillary segments again form a hexagonal network which must have a few interspersed pentagonal meshes, since hexagons can never close a (spheric) surface completely (cf. D'Arcy W. Thompson, 1942, p. 708 ff.). This detail, however, will not trouble us, since we choose to regard the capillary bed as a sheet homogeneously organized into the described meshwork. The network of the entire lung was shown to be subdivided into approximately 200 to 300 million unit networks, each of which receives its blood supply through one precapillary and is drained by one postcapillary. Each of these network units is in contact with several alveoli and, conversely, each alveolus is adjacent to several network units. For the purpose of this model we shall, however, introduce the simplification that the network unit enclosed in its wall is connected to only one precapillary and one postcapillary which we shall place at the left and right poles of the sphere respectively (Fig. 78). It should be noted that the "true" situation is restored if several alveoli are packed closely together with their half-networks merging, as shown in Fig. 79.

The blood enters the simplified model network of Fig. 78 at the arterial pole, moves across the spheric shell, where it is exposed to alveolar air, and leaves the area of the alveolus at the venous pole. It will travel an approximate distance

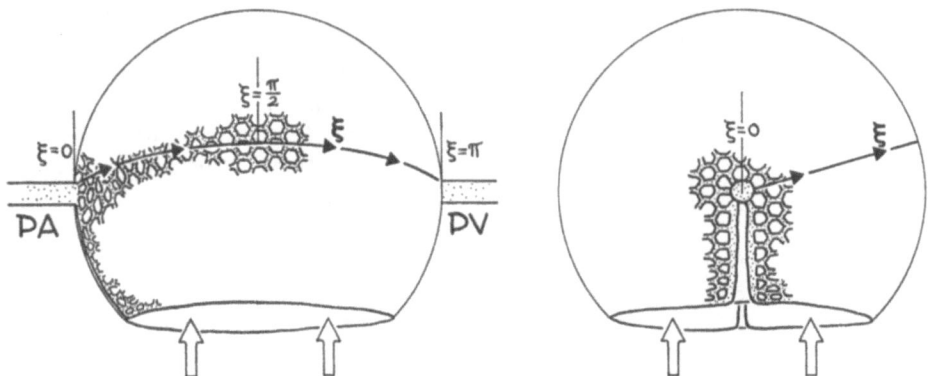

Fig. 78. Model of capillary network in spheric shell of model alveolus described in relation to precapillary (PA) and postcapillary (PV). Compare text

of $(\pi \cdot r)$ from inflow to outflow point. Along this path we can define its location by its distance $w = \xi \cdot r$ from the arterial pole, so that evidently $0 \leq \xi \leq \pi$

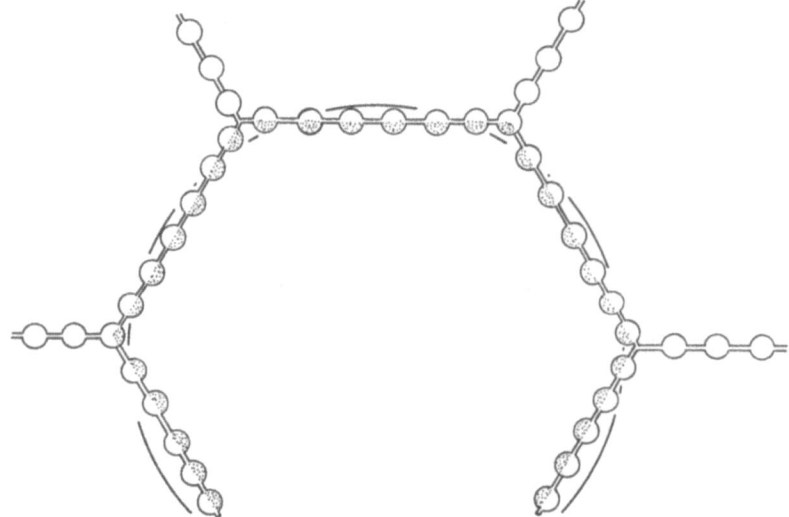

Fig. 79. Arrangement of capillaries in interalveolar septa of model alveoli after "packing"

(Fig. 78); ξ is measured along meridians. Equidistant points will lie on circles or circular arcs of length $C(\xi)$, whereby

$$C(\xi) = \begin{cases} 2\pi r \cdot \sin \xi & \text{if } 0.233\pi \geq \xi \geq 0.767\pi \\ 2\pi r \cdot \sin \xi \left(1 - \cos^{-1}\left[\dfrac{2}{3 \sin \xi}\right]\right) & \text{if } 0.233\pi < \xi < 0.767\pi \end{cases} \quad (9.7)$$

The first equation applies in the regions around the poles, the second holds where the arcs do not encircle the sphere but are interrupted by the entrance plane of the alveolus. In order to have the arc $C(\xi)$ interrupted in both regions in similar fashion, imagine a very narrow wedge-shaped slit cut into the shell from each pole down to the edge of the entrance plane (Fig. 78b). The capillary network is thus delineated by a closed borderline which runs on either side from arterial to venous pole. In Fig. 80 this capillary network was spread out flat, maintaining at each position ξ the proper length of arc $C(\xi)$.

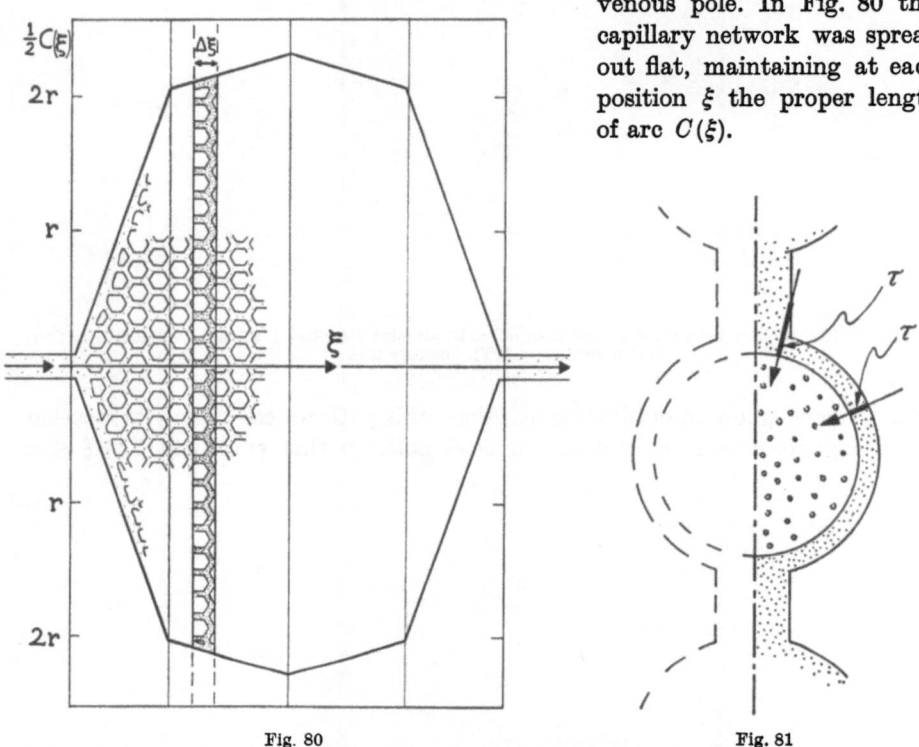

Fig. 80   Fig. 81

Fig. 80. Capillary network of model alveolus spread out flat to demonstrate change in total cross section, etc., as function of distance from the precapillary. Compare text

Fig. 81. Approximate model of alveolo-capillary air-blood barrier. Compare text

As previously stated, the capillaries are homogeneously distributed over the entire alveolus. A zone of length $C(\xi)$ and width $r \cdot \Delta\xi$ will, therefore, contain an average number of capillary segments

$$n_c(\xi, \Delta\xi) = C(\xi) \cdot \Delta\xi \frac{540}{\pi r} \tag{9.8}$$

whereby $r \cdot \Delta\xi \geqq L_c$, L_c being the average length of the capillary segments. Ascribing to the capillary segments a constant average radius R_c and a constant length L_c, their volume v_c and surface s_c can be calculated by equations (7.1) and (7.2); their cross sectional area is evidently $a_c = \pi R_c^2$. Thus the over-all cross section A_c of the capillary network at any position ξ is

$$A_c(\xi) = C(\xi) \cdot \frac{540}{\pi r^2} L_c \cdot a_c . \tag{9.9}$$

The total volume and surface of the capillary network in the zone between ξ and $(\xi + \Delta \xi)$ are

$$V_c(\xi) = n_c(\xi) \cdot v_c \tag{9.10}$$
$$S_c(\xi) = n_c(\xi) \cdot s_c \tag{9.11}$$

whereby only 1/2 of V_c and S_c are really participating in the gas exchange with the given alveolus. The air-tissue interface I_{A-T} and the tissue-blood interface I_{T-B} across which the gas exchange takes place are

$$I_{A-T}(\xi) = 1.35 \cdot C(\xi) \cdot r\Delta\xi \tag{9.12}$$
$$I_{T-B}(\xi) = 0.55 \cdot S_c(\xi) \tag{9.13}$$

[cf. equations (9.2) and (7.14)].

Alveolar air and capillary blood are separated by a layer of tissue which has the properties described in Chapter VIII. As indicated in Fig. 81, this tissue covers about 80% of the cylindric surface of the capillaries as a layer of equal thickness of some 0.45 μ. The remaining 20% of the capillary surface is in contact with the intercapillary tissue membrane. In this region the radial air-blood distance increases gradually from 0.45 μ to infinity for the deepest regions. This arrangement simulates the distribution of thickness which was discussed on p. 104. The harmonic mean thickness of this barrier model is $\tau_h = 0.54$ μ.

So far, the model of the respiratory zone of the human lung has been described in general terms. Its dimensions were determined by the average alveolar volume $\bar{v}_A$ [equation (9.1)], or by the corresponding average alveolar radius r. We can now, introduce specific numerical values, determined in the preceding chapters, into the corresponding equations. Table IX.1 lists a collection of dimensions of a typical model alveolus; it refers to an average adult human lung in a stationary state of about 3/4 maximal inflation and with an average degree of vascular injection (cf. p. 44 and 86).

Table IX.1. *Typical Dimensions of Model of Respiratory Zone for Average Adult*

Degree of inflation of lung	~ 3/4 T. L. C.
Total air volume of lung	4800 ml
Volume of respiratory zone	3150 ml
Total number of alveoli	$300 \cdot 10^6$
Area of air-tissue interface	81 m²
Radius of model alveolus	$1.4 \cdot 10^{-2}$ cm
Depth of model alveolus	$2.3 \cdot 10^{-2}$ cm
Volume of model alveolus	$1.05 \cdot 10^{-5}$ ml
Air-tissue interface of alveolus	$27 \cdot 10^{-4}$ cm²
Total capillary blood volume	140 ml
Total capillary surface of lung	70 m²
Number of capillary segments	$280 \cdot 10^9$
Volume of blood per alveolus	$4.7 \cdot 10^{-7}$ ml
Capillary surface per alveolus	$23.4 \cdot 10^{-4}$ cm²
Radius of capillary segment	$4 \cdot 10^{-4}$ cm
Length of capillary segment	$10.3 \cdot 10^{-4}$ cm
Effective barrier thickness	$0.54 \cdot 10^{-4}$ cm

It would be desirable to keep this model adaptable to varying functional states. But, since we lack information on the actual effect of different degrees of inflation or capillary injection on geometry and dimensions of alveoli and capillaries, this can be done only in a hypothetic sense although a recent study by

STOREY and STAUB (1962) has given this approach some justification. Within certain limits, the alveolar diameter d_A, defined on p. 72 in equation (6.13), can indicate the variation of the radius r_m of our model with varying degrees of inflation ζ (Fig. 45). The degree of vascular injection will influence the average capillary radius R_c, as discussed on p. 85. However, these hypothetic relations should be used with caution; they must be justified experimentally before they can be used with confidence within wide ranges of variation. This was extensively discussed on p. 87.

Chapter X

Geometry and Dimensions of Airways of Conductive and Transitory Zones

The airways of the conductive and transitory zone (cf. p. 9, Figs. 4 and 6) act as distributors of air among the 300 million alveoli of the respiratory zone. To serve this function they are organized into a tree-like system which extends from the trachea into the respiratory portion of the lung. It is the purpose of this chapter to study the pattern of organization of the airways, and to determine the dimensions of its elements.

A. Architecture of the Human Airway System

1. General Organization and Definition of Terms

In Fig. 82 the organization of the airways has been diagrammatically summarized. Starting with a single trunk, the trachea, the number of airways multiplies throughout the conductive and transitory zone by a pattern of branching which will be shown to conform with the basic pattern of dichotomy. Toward the periphery alveoli will gradually become apposed to the lateral wall of the transitory airways (cf. p. 56, Figs. 5 and 38).

The element of the conductive and transitory airways is, therefore, defined as the unit airway, stretching from the termination of its parent to the origin of its two daughter airways (Fig. 5). The two elements branching from the same parent will be called conjugates. As we progress from the trachea toward the periphery, the wall assumes different structural features characterizing the following types of elements (cf. MILLER, 1937; VON HAYEK, 1953):

a) *Bronchi*, in the wall of which "passive" tissue components prevail. Their fibrocartilaginous coat encloses rings or plates of cartilage in a strong layer of collagenous and elastic fibrous tissue. A relatively thin smooth muscle layer underlies the thick epithelial structures.

b) *Bronchioles*, in the wall of which is contained a relatively large amount of "active" smooth muscle, while the fibrous coat is rather loose and thin, with no cartilage as solid support.

c) *Respiratory bronchioles*, showing a wall structure identical with that of ordinary bronchioles, but interrupted where single or small groups of alveoli are apposed to the lumen (partial alveolation).

d) *Alveolar ducts*, which may be regarded as "respiratory bronchioles", in which the entire wall has been occupied by alveoli (complete alveolation). Of the wall structure only a fine meshwork of elastic and thin collagenous fibers with few single smooth muscle fibers remains. This meshwork forms the "entrance rings" to the alveoli.

e) *Alveolar sacs*, which are essentially identical in structure with alveolar ducts. At their peripheral end, however, they do not branch further but are closed off by a group of terminal alveoli.

2. Pattern of Airway Branching

The introductory statement that airways multiply toward the periphery by dichotomy or "branching into two" requires further scrutiny. We can distinguish between two types of dichotomy, schematically illustrated in Fig. 83. In the case of regular dichotomy, the two conjugate elements are of equal dimensions and

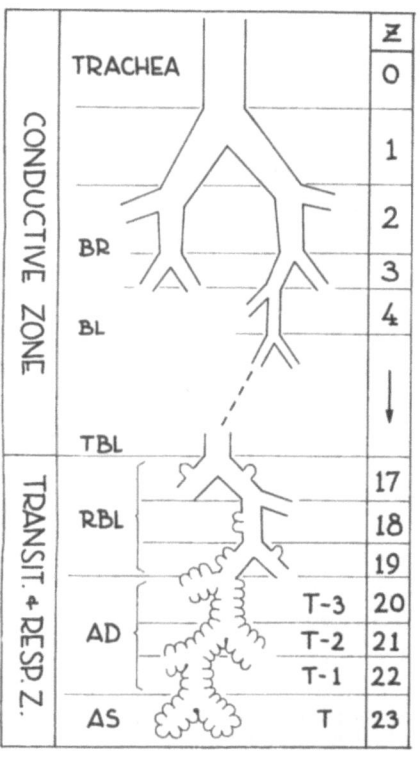

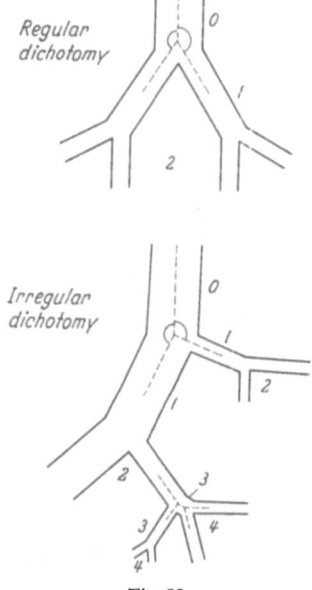

Fig. 82

Fig. 83

Fig. 82. Diagrammatic representation of the sequence of elements in the conductive and transitory zones of the airways. z designates the order of generation of branching, T the terminal generation. Compare text

Fig. 83. Pattern of regular and irregular dichotomy with apparent monopody shown in first, and apparent trichotomy in third generation

branch off their parent at equal angles. In a strict sense this situation is quite improbable in nature. Mostly we shall find some degree of irregularity of dimensions which leads to the pattern of irregular dichotomy, where lengths or diameters of the two conjugate elements are more or less unequal (Fig. 83b). It has also been shown — for example, for blood vessels (HESS, 1915) — that the angle at which a branch comes off its parent is proportional to the ratio of the dimensions of parent and daughter vessel: A small branch takes off almost at right angles, while the direction of a large branch may deviate only slightly from the direction of its parent. In irregular dichotomy we shall, therefore, also observe a variation in the branching angles. If the diameters of the conjugate elements differ widely, the larger conjugate may not be distinguishable from its parent, so that the small

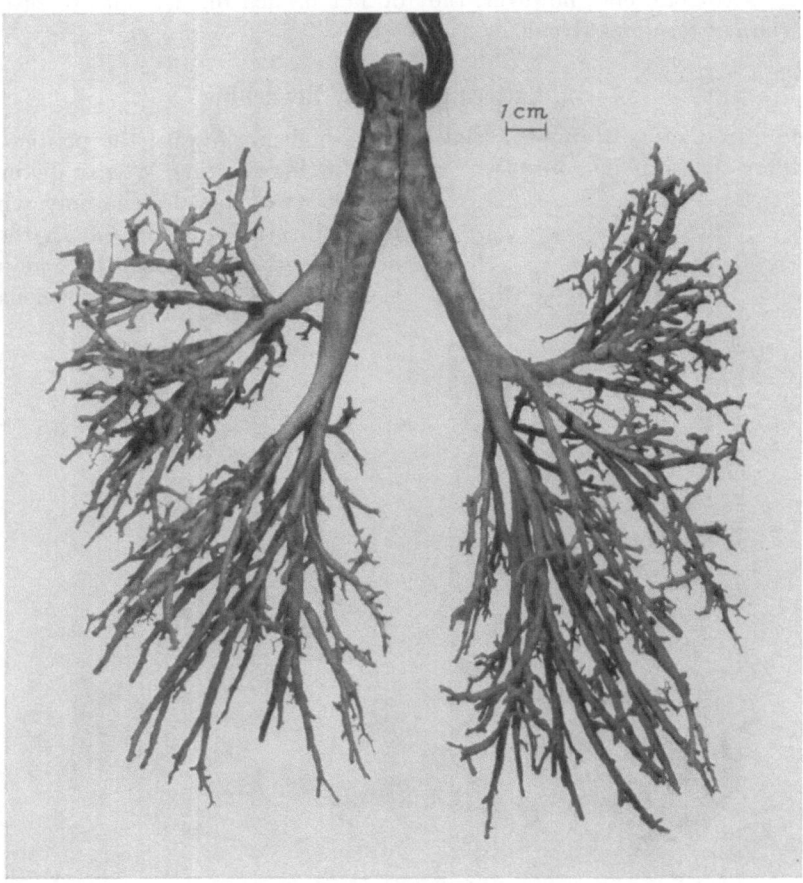

Fig. 84. Anterior view of plastic cast of human bronchial tree of an adult used for measurement of airway dimensions in first 10 to 15 generations. 1:2. Courtesy of Dr. AVERILL A. LIEBOW, Yale University School of Medicine

branch appears to originate laterally on a large stem. This extreme case of irregular dichotomy approaches what is called monopody (Fig. 83 b). On the other hand, if one of the conjugate elements is very short, it may become inconspicuous, so that its two daughter branches appear to be conjugated with its own conjugate, thus simulating trichotomy (Fig. 83 b). Monopody and trichotomy in themselves exist in nature (cf. D'ARCY THOMPSON, 1942). Several botanic examples could be given. However, if they occur in a system which, in the vast majority of branchings, follows the pattern of irregular dichotomy, we may regard them as merely apparent, representing the extreme cases of irregularity.

It has long been known (EWART, 1889; ROHRER, 1915; BENDER, 1923; POLI-
CARD et GALY, 1945; VON HAYEK, 1953; REID, 1958; and others) that the airways
of the human lung branch by irregular dichotomy. This is immediately realized
upon examination of the casts in Fig. 84 and Fig. 2. Most of those branchings,
which lie parallel to the plane of view, will appear to be obvious dichotomies.
It is also evident that the irregularity is mostly due to a strong variation in length
rather than to differences in diameters, as will be discussed later. Apparent

monopodies are, therefore, rare in the
human lung, although various authors
have claimed that major bronchi branch
monopodically (AEBY, 1880; MILLER,
1937). This may be different with lungs
of other mammals. Apparent trichotomy,
however, is not too uncommon. One well-
known example is the apparent division
of the right upper lobe bronchus into
three segmental bronchi. Upon careful
examination, however, a short bronchus
can be seen joining two of the three
segmental bronchi; in the cast of Fig. 84,
the apical and anterior segmental bronchi
had a common stem.

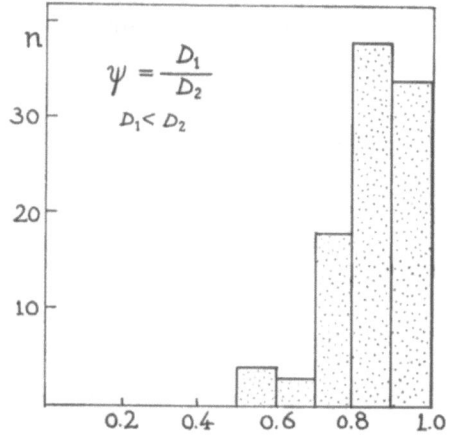

As will be described, a binary numbering system was used to identify the
bronchi of cast 1. It was possible to
apply this code without difficulty to all
of the more than 1,000 branchings analyzed. This demonstrates convincingly
that the basic pattern of branching of the
human bronchial tree is really that of irregular dichotomy, as already defined. It
should, however, be noted that alveoli
do not branch off the alveolar ducts by
dichotomy.

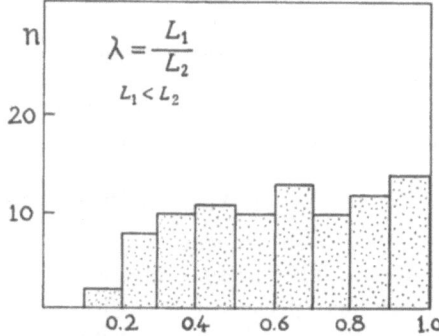

Fig. 85. Distribution of ratio ψ of smaller to larger
diameter, and of ratio λ of shorter to longer length of
conjugate airway segments

The degree of irregularity of dichotomy
could be assessed by determining the
ratio of the lengths of the two conjugate
elements, and the ratio of their diameters. For this purpose the shorter dimension
was divided by the longer one, yielding the coefficients

$$\lambda = \frac{L_1}{L_2} \leq 1, \quad \text{where} \quad L_1 < L_2, \tag{10.1}$$

$$\psi = \frac{D_1}{D_2} \leq 1, \quad \text{where} \quad D_1 < D_2. \tag{10.2}$$

Figure 85 shows the distribution of λ and ψ as found for generations 5 to 7;
λ varied from 0.14 to 1.0, and averaged 0.62 with a standard deviation of ± 0.20.
For the same generations ψ ranged from 0.5 to 1.0 with an average of 0.86 ± 0.01.

In 8% of the cases the length, and in 35% the diameter, of the conjugates was equal. The irregularity of branching is, therefore, caused rather by variations in length than in diameter.

3. Geometry of the Airway Elements

The airway element has approximately the shape of a hollow cylinder. Its two ends are trimmed in wedge shape so as to allow junction with other elements; at its proximal end it is in connection with its parent and its own conjugate (Fig. 86), while at its peripheral end it communicates with its two daughter branches. Such geometric bodies have been described for the element of the alveolar capillary network (p. 76).

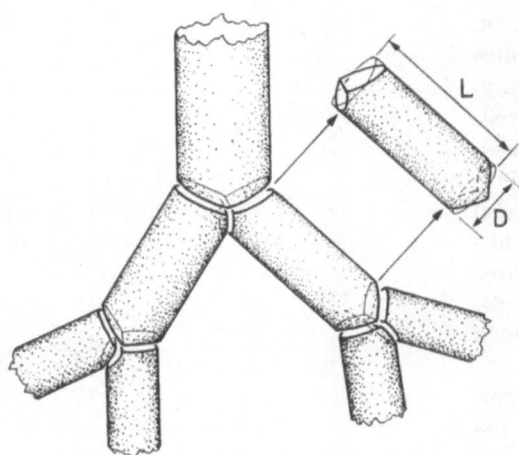

It will be shown that the diameter of the daughterb ran-ches is smaller than that of the parent. Therefore, bronchi with the shape of circular cylinders cannot bejoined without leakage, unless their wall is distorted toward the ends. In a system of regular dichotomy a good fit can be obtained with elliptic cylinders or with slightly conic elements. In irregular dichotomy the shape must be more complex. The re-lative crudeness of the dimensions obtained subsequently, however, does not justify the use of more complicated models in the further analysis.

Fig. 86. Diagram of simplified shape of airway elements

4. Analytic Properties of a System of Dichotomies

For our subsequent analysis of the airways two properties of dichotomous systems will be useful. Since every daughter branch in turn becomes the parent of two more peripheral branches, we can consider the entire system as a sequence of generations of branching, whose order will be denoted by "z". The trachea (the only element which does not have a conjugate) would correspond to generation $z = 0$, the two main bronchi to generation $z = 1$, and so on. This is illustrated in Fig. 87 for the first four generations of the cast of Fig. 84.

Since every element located in generation $(z - 1)$ gives origin to two branches which will appertain to the subsequent generation z, it follows that the total number of branches $n(z)$ in generation z will be

$$N(z) = 2 \cdot n(z - 1) = 2^z .\tag{10.3}$$

It follows, furthermore, that the total number of branches $N_{(z)}$ from the origin to generation z inclusively will be

$$n(z) = \sum_{s=0}^{z} 2^s = 2^{z+1} - 1 .\tag{10.4}$$

The definition of a sequence of generations, together with the above numerical relations between generations, will enable us to devise a procedure of systematic sampling (see below, which will overcome some of the difficulties arising from

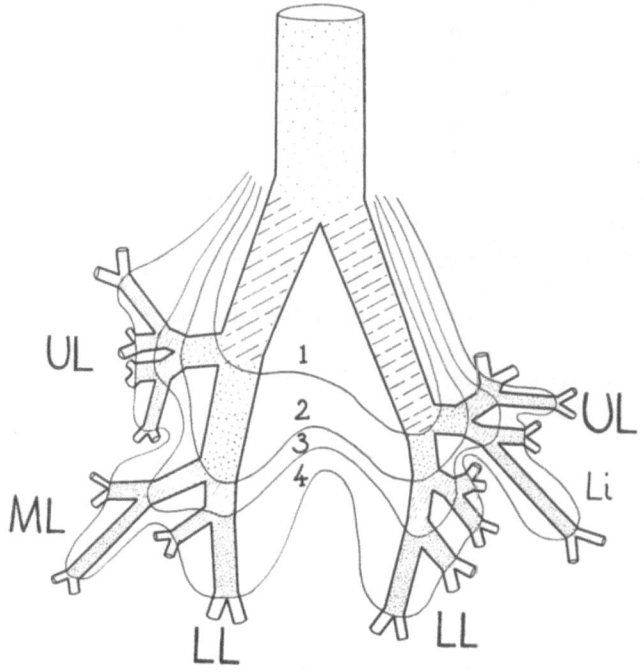

UL

UL

ML

Li

LL   LL

Fig. 87. First four generations of airways of the cast redrawn to scale in two dimensions

non-random distribution of the conductive airways. The relations (10.3) and (10.4) will also be used in setting up a model of the airway system (see p. 136).

B. Dimensions of Proximal Airway Elements of the Conductive Zone

The polar orientation of the major airway elements of the conductive zone deprives us of the possibility of using random sampling procedures in analyzing their dimensions, mainly because the systematic centrifugal change in dimensions should be reflected in the data obtained (see p. 123). Since airways were found to branch by dichotomy, it was possible to design a systematic sampling procedure by stratification of the airway elements with respect to generations of branching. Within each generation "complete" sampling of all elements was attempted, since obtaining significant random samples appeared to be impossible. In those generations in which "completeness" could not be achieved, the sample obtained was affected by a positive bias, since the missing elements belonged to the classes with small dimensions. By means of an appropriate statistical treatment, however, it was possible to correct this bias (see p. 118).

1. Method of Analysis

To arrive at an estimate of the variation of diameter and length of the bronchi of the first 10 generations, a plastic replica of the cavity of the bronchi of an adult human lung was employed. This cast, illustrated in Fig. 84, was prepared by Dr. AVERILL A. LIEBOW of Yale University, according to his well-known technique. It is a partial filling of the bronchial tree, down to branches of about 1.5 mm in diameter, in a state of moderate expansion which probably compares well with the degree of inflation used for preparation of the specimens for microscopic studies, as previously described. We would estimate that this lung would have had a lung volume (at 3/4 TLC) of 5 to 6 liters if prepared by our procedure (cf. p. 41).

In order to fulfill the requirements of this study, the airways of this specimen were analyzed through the following steps:

a) Each bronchial segment had to be labelled with a code indicating its location with respect to its origin and to its generation. This cast was, therefore, photographed from various angles. On large prints and tracings of detailed regions each segment was identified and labelled. Since we were analyzing a system of dichotomous branching, a binary numbering system provided an appropriate code. Each pair of two conjugate branches originating from the parent was, therefore, given the letters a and b. Starting with A for the trachea, the code for the branches in generations 0 to 3 . . ., would, for example, be

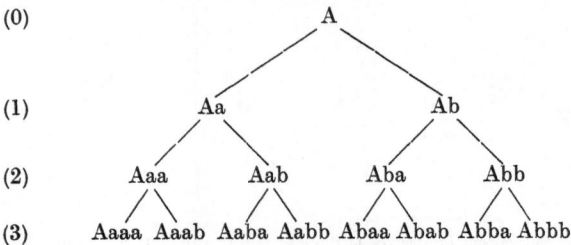

For practical handling some abbreviations were introduced whenever the code became too long, but this did not affect the basic concept.

b) Midpoint diameter and length of each segment were measured with a divider (caliper) and recorded alongside the code on a pedigree chart. This made a variety of surveys possible.

c) Break-off points were also recorded, as well as occasionally present distortions of the cast, which were not measured.

2. Results of Measurement

The dimensions of the bronchial segments, as measured on this cast, are plotted in Fig. 88 for each generation. The diagram relates diameter to length, each dot representing one segment. At first sight there is no obvious correlation between L and D. The calculation of the average length $\bar{L}_D$ of the segments for each diameter class, however, shows that $\bar{L}_D$ (circles) follows approximately a straight line through the origin. The slope of this line $1/\delta_z$ represents the average length-to-diameter ratio for each generation, as will shortly be discussed.

The diagrams of Fig. 88 reveal also that the range of variation in length of the bronchial segments greatly exceeds that of the diameters. The frequency distribution of diameters and lengths in each generation are plotted as histograms in Figs. 89 and 90, respectively. It can be seen that the most frequent diameter shifts toward smaller values with increasing order of the generations, the histograms demonstrating also an increasing skewness to the right. The histograms of length are broader. They appear highly skewed to the right.

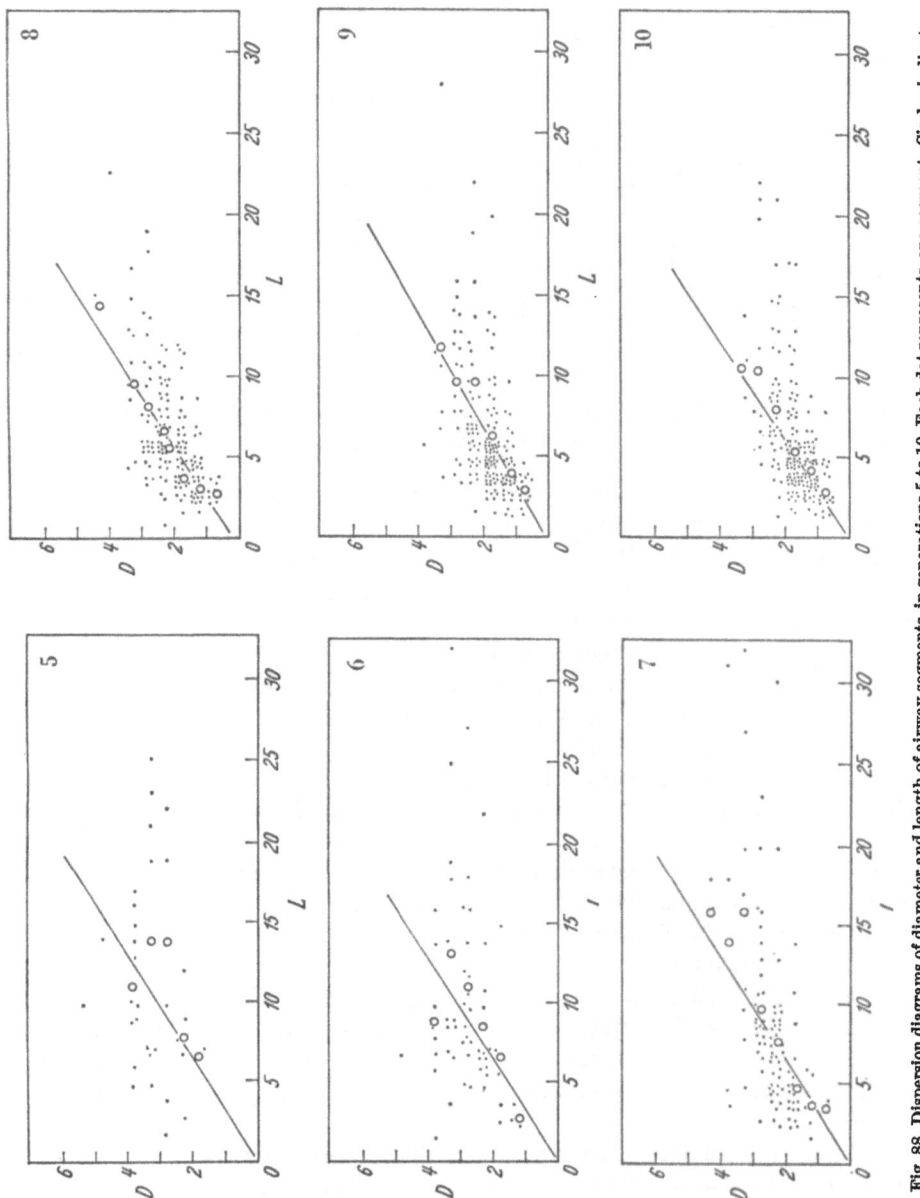

Fig. 88. Dispersion diagrams of diameter and length of airway segments in generations 5 to 10. Each dot represents one segment. Circles indicate average segmental length in each diameter class. Slope of diagonal corresponds to average diameter-to-length ratio in each generation

No averages have been calculated from these frequency distributions, since they are most probably not representative samples of the entire population of branches in each generation. It is technically not possible to measure bronchial dimensions in a cast that is "complete". Therefore, in our specimen, a large number of *small* branches were broken off or not filled. It was possible, however, to estimate, in each generation, how many branches were missing, since the dichotomous system of branching allots a given number $n = 2^z$ of elements to each generation z (see above). In Table X.1 the number of branches to be expected in each generation and the numbers measured for D and L are indicated. Generations 0 to 5 were completely sampled while in the higher orders the fraction of the total stratum population represented in the sample decreased gradually. The approximately 200 branches measured in the tenth generation represent only about 20% of the total.

Table X.1. *Sampling Fractions of Bronchial Cast*

Generation z	Number of branches predicted N_z	Diameter measured		Length measured	
		N_D	N_D/N_z	N_L	N_L/N_z
0	1	1	1.0	1	1.0
1	2	2	1.0	2	1.0
2	4	4	1.0	4	1.0
3	8	8	1.0	8	1.0
4	16	16	1.0	16	1.0
5	32	32	1.0	32	1.0
6	64	61	0.95	58	0.91
7	128	109	0.85	103	0.80
8	256	176	0.69	153	0.60
9	512	214	0.42	181	0.35
10	1024	216	0.21	188	0.18

3. Completion of Distribution of Diameters

In order to reach a reliable estimate of the "true" average dimensions of the bronchial segments, the distribution of sizes had to be completed. It is obvious that the broken-off branches were not randomly distributed over the entire range of diameters, so that a larger fraction had to be added to the smaller sizes than to the larger ones. This correction was based on the following reasoning:

a) The record of break-off points revealed that no branches with diameter $D \geqq 2.5$ mm were broken off. We can, therefore, conclude that the probability of having measured *all* branches with diameter $D \geqq 2.5$ mm is almost 1 for all generations. The probability of having measured all branches of diameter $D < 2.5$ mm decreases with the diameter and is zero for diameters $D < 1.0$ mm since no branches smaller than 1 mm could be recorded.

b) It will later be shown that the diameter of the most peripheral airway branches is approximately 0.4 mm. This sets the lower limit for the airway diameter to be in the class of 0.5 mm.

c) It is assumed that the "true" distribution of diameters can be approximated by a binomial distribution over the range of classes for $0.5 \leqq D \leqq D_{max}$, where D_{max} is the largest diameter recorded in each generation. In support of this assumption we observe that the diameter histograms of the two nearly complete generations 6 and 7 appear well simulated by a corresponding binomial distribution (Fig. 89).

On the basis of these three premises the following procedure was used to determine the binomial distribution of diameters in each generation:

The diameters were grouped in classes of interval 0.5 mm. The frequency of occurrence $F(x)$ of a given class $x = \dfrac{D}{0.5}$ is by the binomial function

$$F(x) = \frac{m!}{(m-x)! \cdot x!} \cdot p^{m-x} \cdot (1-p)^x \qquad (10.5)$$

where

$$m = \frac{D_{max} - 0.5}{0.5} \qquad (10.6)$$

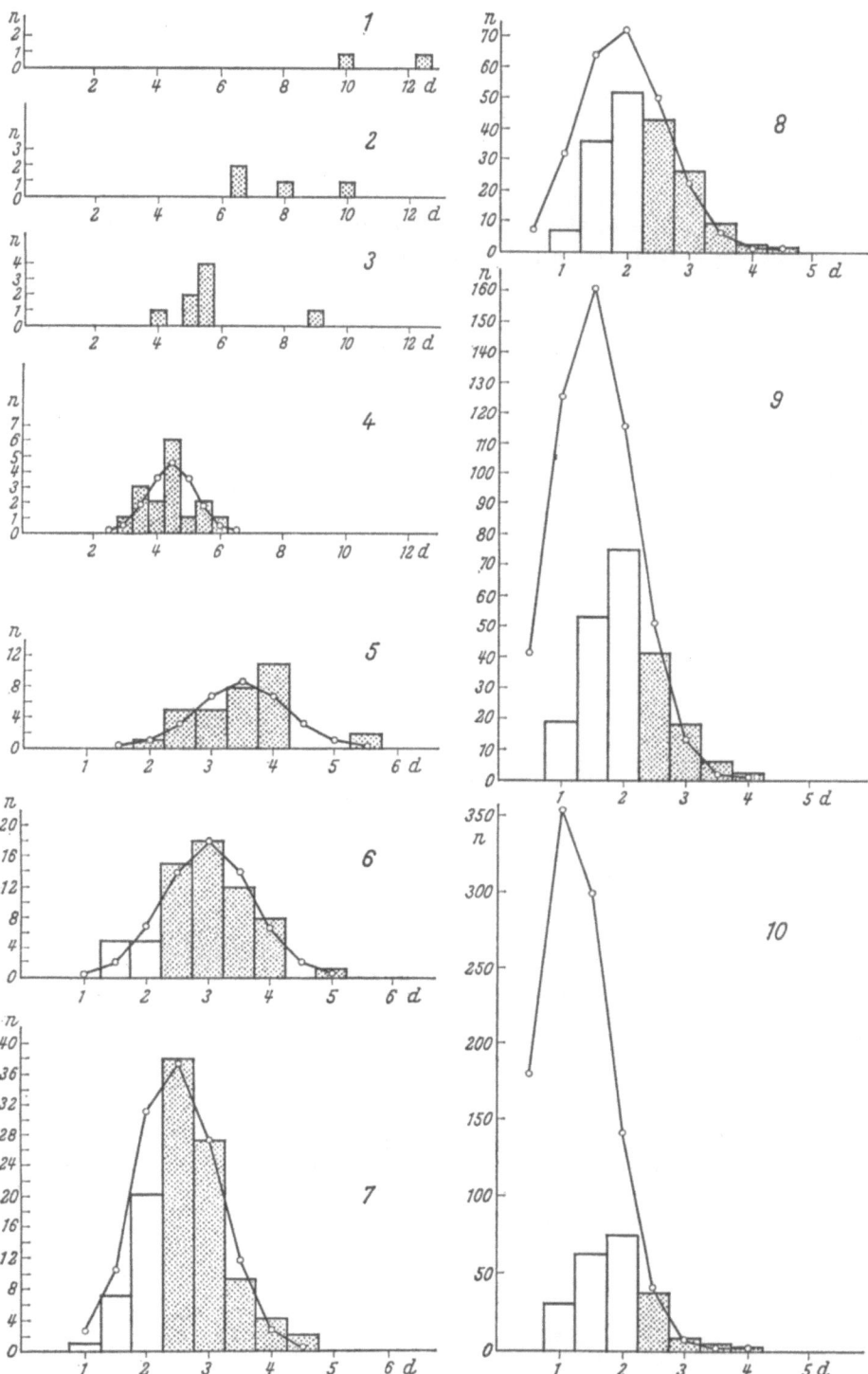

Fig. 89. Distribution of airway diameters in generations 1 to 10. Histograms represent measurements on castof Fig. 84. Those classes which were completely sampled are stippled. Superimposed curves are binomial functions derived on the basis of the data to complete the partially sampled classes. Compare text

This choice of m satisfies the condition that the distribution should range from 0.5 to D_{max}.

To satisfy the premise that all branches with diameter $D \geq 2.5$ had been measured, the coefficient p was chosen so that

$$\sum_{D=2.5}^{D_{max}} F\left(\frac{D}{0.5}\right) \cdot n(z) = \sum_{D=2.5}^{D_{max}} n'(D) \tag{10.7}$$

where $n(z) = 2^z$ is the total number of branches in each complete generation [from equation (10.3)], and $n'(D)$ is the number of branches with diameter D

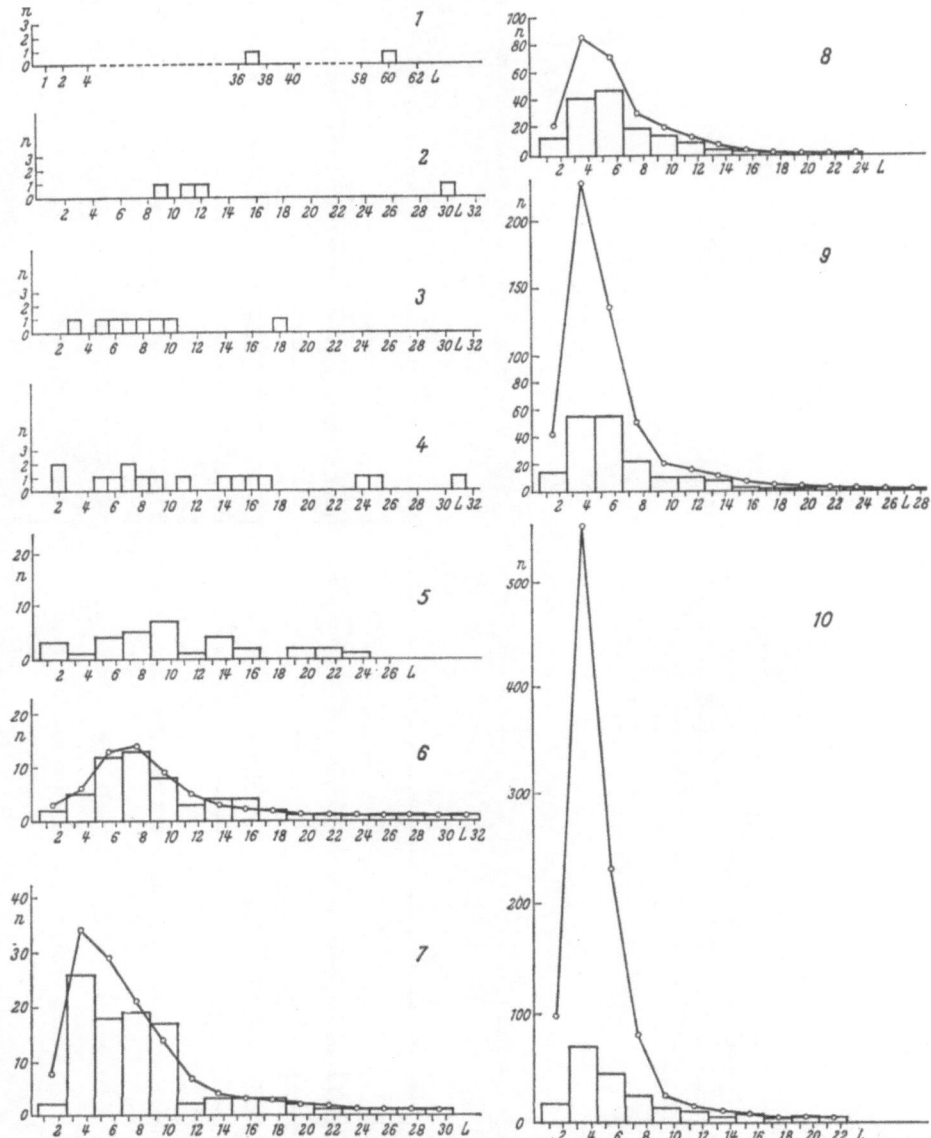

Fig. 90. Distribution of lengths of airway segments in generations 1 to 10. Histograms represent measurements obtained on cast of Fig. 84. Superimposed curves indicate "completed" distributions. Compare text

measured on the cast. This assures that the number of branches larger than 2 mm in the reconstructed "true" distribution is the same as that actually measured.

These conditions define a unique binomial distribution for each generation. They have been plotted over the histograms in Fig. 89. Thus, the "true" average diameter of the airways in each generation could be calculated from the binomial distributions for the incompletely sampled generations 6—10, while the actual data were taken for the first five generations. The average diameter was found to be a decreasing function of the generation, as will be explained in detail.

4. Completion of the Distribution of Lengths

The distribution of length of the bronchial segments in each generation could not be completed for the missing branches by a direct procedure as was possible for the diameters. The following reasoning, however, appeared to be justified:

a) The missing lengths belonged to those branches whose diameter was also missing; thus, the number of branches of $D < 2.5$ mm added to the histograms by the binomial distribution had to form the basis for the completion of the length distribution.

b) The number $n_z(D_+)$ of branches added to each diameter class D is

$$n_z(D_+) = F\left(\frac{D}{0.5}\right) \cdot n_{(z)} - n'_D \quad (10.8)$$

where $F\left(\frac{D}{0.5}\right) n_{(z)}$ is the "complete", n'_D the measured number of branches in the class D. $n_z(D_+)$ had to be distributed over the length classes according to a distribution pattern which could be determined from those branches which were measured. Figure 91 presents these distributions for $D = 2, 1.5$ and 1. They were obtained by pooling all branches of a given diameter from generations 7 to 10. These curves yielded the frequencies

$$f_D(L) = \frac{n_D(L)}{\sum\limits_L n_D(L)} \quad (10.9)$$

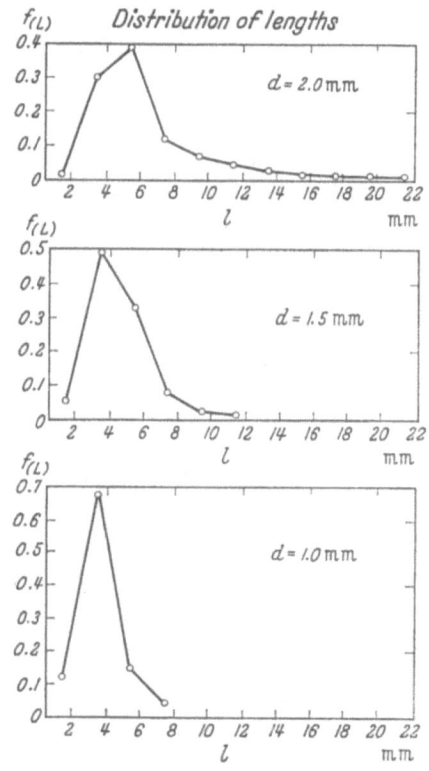

Fig. 91. Distribution of lengths of airway segments in diameter classes 2, 1.5 and 1 mm

where $n_D(L)$ is the number of branches with length L observed at diameter D. A corresponding distribution for $D = 0.5$ was not available. It was, therefore, assumed that the distribution of $D = 1$ would represent a fair approximation.

The completion of the distribution of lengths was, accordingly carried out by the following steps:

The number of branches $n_z(D_+)$ added to each diameter class D was distributed over the length classes L by the function,

$$n_z(L_+) = \sum_D f_D(L) \cdot n_z(D_+) \tag{10.10}$$

where $n_z(L_+)$ is the number of branches to be added to each length class L. The "complete" number of branches in each length class is, therefore,

$$n(L) = n'_L + n_z(L_+) \tag{10.11}$$

where n'_L is the number of branches actually measured.

A small additional correction of the order of 5 to 10% had to be made because a few branches could be measured for diameter but not for length (see Table X.1).

The completed distributions of lengths are again plotted over the histograms in Fig. 90. They show that the increase in the number of branches was distributed over almost the entire range. The shorter branches were, however, greatly favored. This appears reasonable in view of the assumption that the average length-to-diameter ratio is about the same for all diameters (Fig. 92 b), and that mostly branches of small diameter were missing.

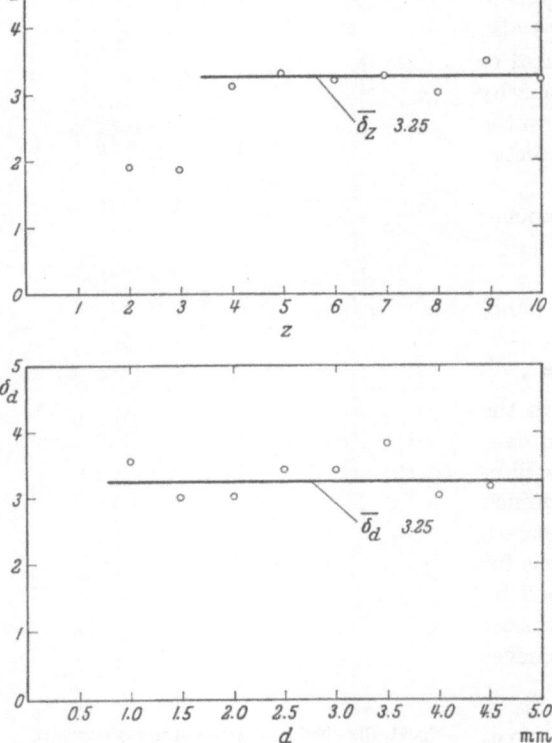

Fig. 92. Average length-to-diameter ratio δ of airway segments.
a per generations z, *b* per diameter class d

5. Average Length-to-diameter Ratio

The dispersion diagrams of Fig. 88 indicate that the length-to-diameter ratio of individual airway elements $\delta = \dfrac{L}{D}$ will vary over a very wide range. Nevertheless, is was already seen in these diagrams that the average lengths $\bar{L}_D$ in each diameter class followed approximately a straight line with slope $1/\delta_z$ corresponding to the average length-to-diameter ratio in the given generation. A major deviation was only observed in those classes which contained a small number of branches.

In Fig. 92a the average $L-D$-ratio δ_z is plotted against the order of generation z. It shows that for generations 4 to 10 δ_z is essentially constant with an average $\delta_z = 3.25$. The δ_z for generations 1 to 3 sharply deviates from this average, but the small number of branches (2 to 8) in these generations can account for this.

In Fig. 92 b the $L - D$-ratio δ_D is plotted against the corresponding diameter class. The $L - D$-ratios of generations 4 to 10 were pooled and an average for each class was thus calculated. δ_D is again seen to be essentially constant for all diameter classes from 1.0 to 4.5 mm; the greatest deviation amounts to less than 20% of the mean which was again 3.25.

6. Average Dimensions of Airways of First Ten Generations

a) Average Diameters

As we have outlined, the average diameter of airways in each of the first ten generations was calculated from the "completed" binomial distributions. In Fig. 93 these diameters, which were seen to be decreasing with increasing orders of generations, are plotted semilogarithmically against the generations. Since the diameters of the intrapulmonary bronchi (generations 2 to 10) appear to follow approximately a straight line, it is suggested that they decrease by an exponential function. Approximate considerations on the loss of energy by frictional resistance in a dichotomous tubular system (cf. D'ARCY THOMPSON, 1942; p. 954) show that a "good" relation between the diameters of the tubes in succeeding generations is

$$d_z = d_{z-1} \cdot 2^{-\frac{1}{3}}. \quad (10.12)$$

This yields the function

$$d_z = d_0 \cdot 2^{-\frac{z}{3}} \quad (10.13)$$

which relates the diameters of the tubes in generation z to the first diameter d_0.

In Fig. 93 this function was fitted to the average bronchial diameters of generations 2 to 10 (intrapulmonary bronchi) by least square fitting on their

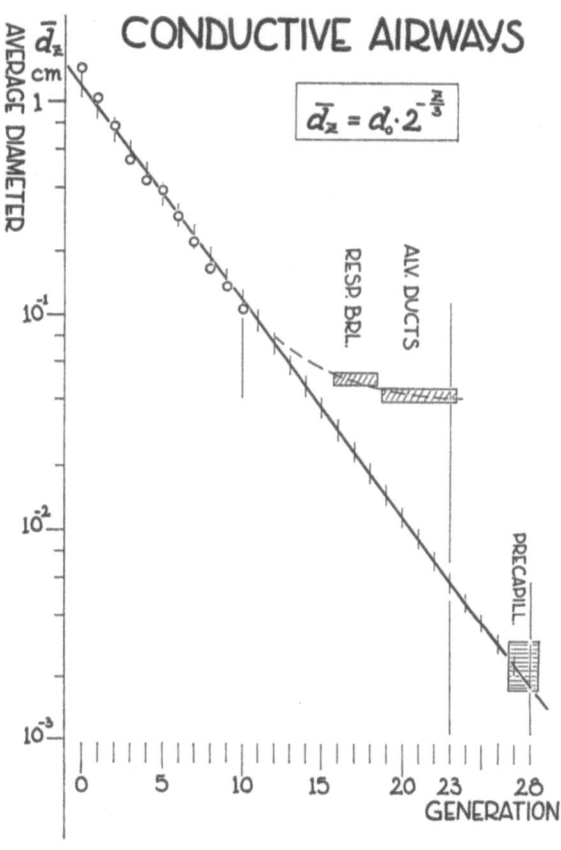

Fig. 93. Average diameter of airways plotted semilogarithmically with respect to generations. Open circles relate to measurements obtained on a cast (completed distributions, cf. text). They appear to follow straight line. Compare text (from WEIBEL and GOMEZ, 1962 b)

logarithms. Diameters of the trachea and main bronchi (generations 0 and 1) were not included in the calculation since they deviate grossly from any possible straight line connecting the other points. The coefficient d_0', therefore, represents the intercept at the origin ($d_0' = 1.2$) rather than the diameter of the trachea

($d_0 = 1.8$). It can be seen that the experimental points follow function (10.13) rather closely, their average deviation from the function being approximately 6%. However, it will be shown on p. 134 that function (10.13) can only describe the average diameters of the major intrapulmonary conducting airways but does not hold toward the periphery, because alveolar ducts were found to be much larger than the extrapolation of equation (10.13) would predict.

b) Average Lengths

For generations 0 to 5 the average length of the airway elements was calculated from the actually measured dimensions, since these could be completely sampled

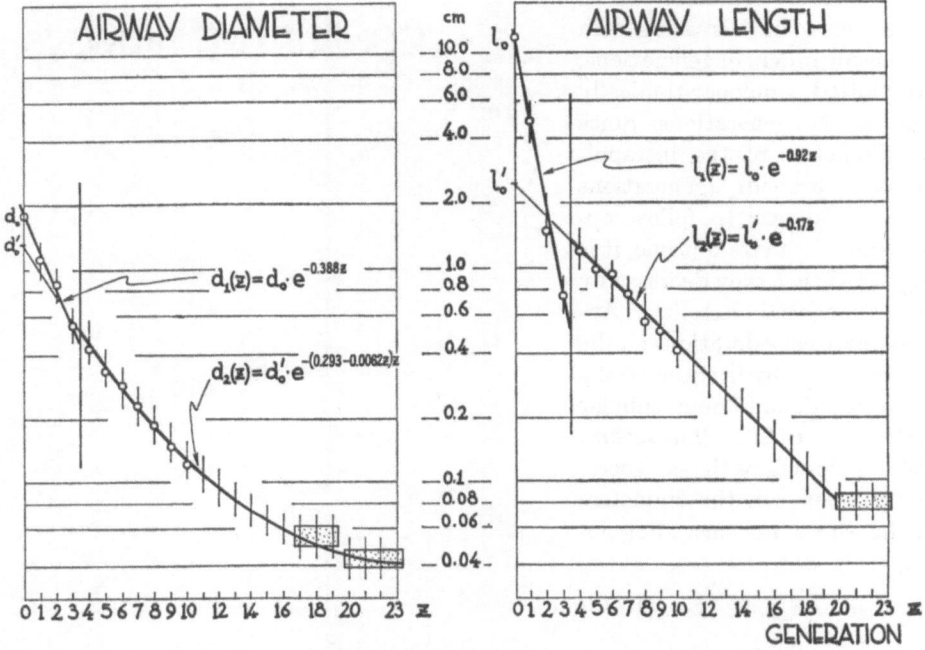

Fig. 94. Average diameter and length of airway segments plotted semilogarithmically against airway generations. Open circles indicate measurements derived from data on cast; stippled boxes signify range of dimensions of alveolar ducts and sacs and of respiratory bronchioles, respectively. Curves relate empiric functions fitted to data. Dimensions refer to adult lung. Compare text

(Table X.1). In order to obtain representative estimates of the average lengths of the incompletely sampled generations 6 to 10, two procedures were used: The average length L_z was first calculated from the "completed" distribution described. Because the completion of these distributions required an involved procedure, a second set of values for the average length was computed by multiplying the average diameter $\bar{D}_z$ (Fig. 93) by the average $L - D$-ratio in that generation (Fig. 92a):

$$L'_z = \delta_z \cdot \bar{D}_z. \tag{10.14}$$

These two estimates of L_z differed from each other by about $\pm 6\%$. In Fig. 94b their averages have been plotted as circles against the generations z. It is evident that the average length of airways for the first 10 generations cannot be described by a continuous function because of a discontinuity observed between the third and fourth generations. It appears, however, that the points lie approximately on two separate straight lines: The length of the trachea $L_0 = 12$ cm and the average length of main bronchi, lobar and segmental bronchi (generations 0 to 3) fall on a steep line which is described by

$$L_1(z) = L_0 \cdot e^{-0.92 \cdot z} . \tag{10.15}$$

The average lengths of the airways in generations 4 to 10 follow approximately a less steep line described by

$$L_2(z) = L_0' \cdot e^{-0.17 z} \tag{10.16}$$

where $L_0' = 2.5$ cm is the intercept at the origin.

7. Variation of Airway Dimensions

The determination of average airway dimensions in successive generations brough out the regular features of the airway system. The following analysis of the data will furnish information on its irregularity. In Figs. 89 and 90 we demonstrated the variation in diameters and lengths of the airway segments in each generation; it would be of interest next to evaluate the influence that this might have on the over-all dimensions of the conducting airway system of the whole lung.

The dimensions of airways decrease consistently with each generation. In a system of regular dichotomy, all branches of a given diameter D^* would be located in the same generation,

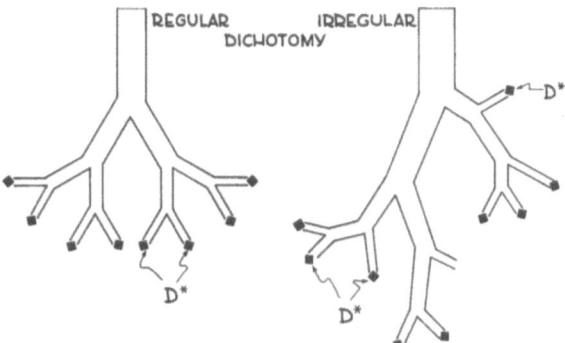

Fig. 95. Diagrammatic representation of distribution of bronchi with diameter D^* in regular and irregular dichotomy

and at the same distance from the origin of the airway system (Fig. 95a). In irregular dichotomy, branches with diameters D^* would be found in various generations, and at different distances from the origin (Fig. 95b), so that the degree of this variation is a measure of irregularity.

Figure 96 shows the distribution of $D^* = 4, 3, 2.5$ and 2 mm, as measured on our cast, over generations z. Branches of 2 mm diameter are found in generations 4 to 13 with highest density around generation 8. With respect to generations the irregularity is therefore considerable. It is less apparent, however, with regard to the distance Δ at which branches of diameter D^* are found (Fig. 97), if Δ is measured from the root of the trachea. But if we study the distance Δ' of branches with $D_2^* = 2$ mm from branches with $D_4^* = 4$ mm (Fig. 98), Δ' will vary appreciably between 0.5 and 10 cm.

Figure 99 shows a photograph of the cast examined in this study from which all branches with diameter $D < D^*$ have been removed; in the right lung $D^* =$

3 mm, and in the left lung $D^* = 2$ mm. The break-off points thus indicate the topographic location of D^*. It is quite evident from this picture that the variation in the distances Δ or Δ' is due to the shape of the lung: D^* is close to the origin in

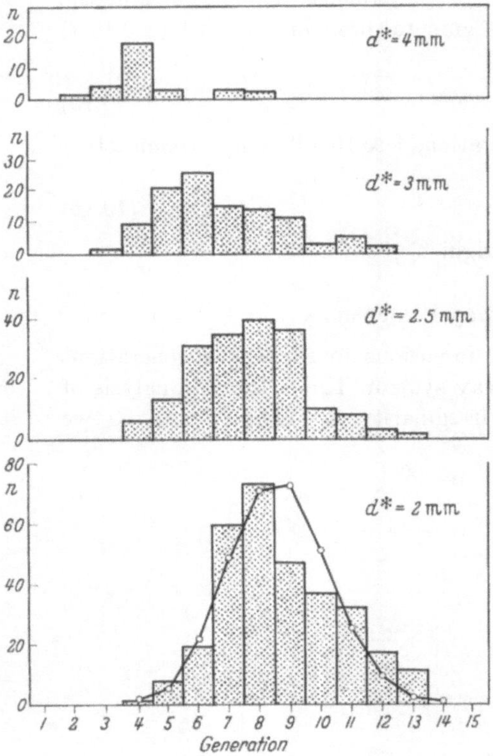

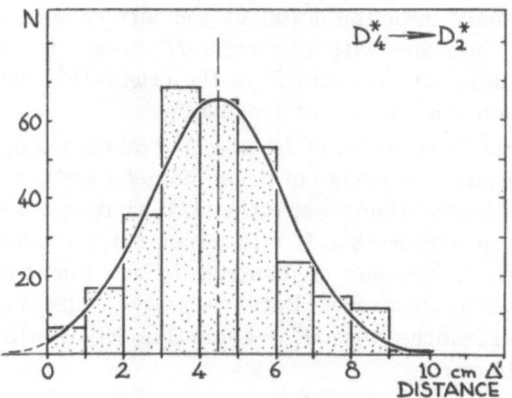

Fig. 96. Distribution of airways of diameter D^* per generations as determined on cast. Curve represents binomial function fitted to data

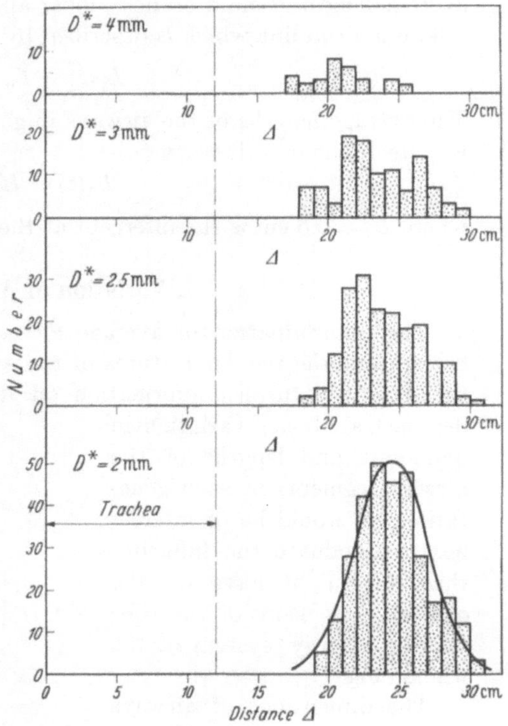

Fig. 97. Distribution of airways of diameter D^* per distance Δ from the origin of the trachea. Curve in lower graph is normal distribution fitted to data

Fig. 28. Distribution of distances Δ' of branches of $D_2^* = 2$ mm from their precursors of $D_4^* = 4$ mm. Approximation of distribution by normal curve with small error. Compare text

central regions, while it is farthest removed in the peripheral regions of the lower lobes. It is evident also that the shape of the lung becomes better approximated the smaller the value assigned to D^*.

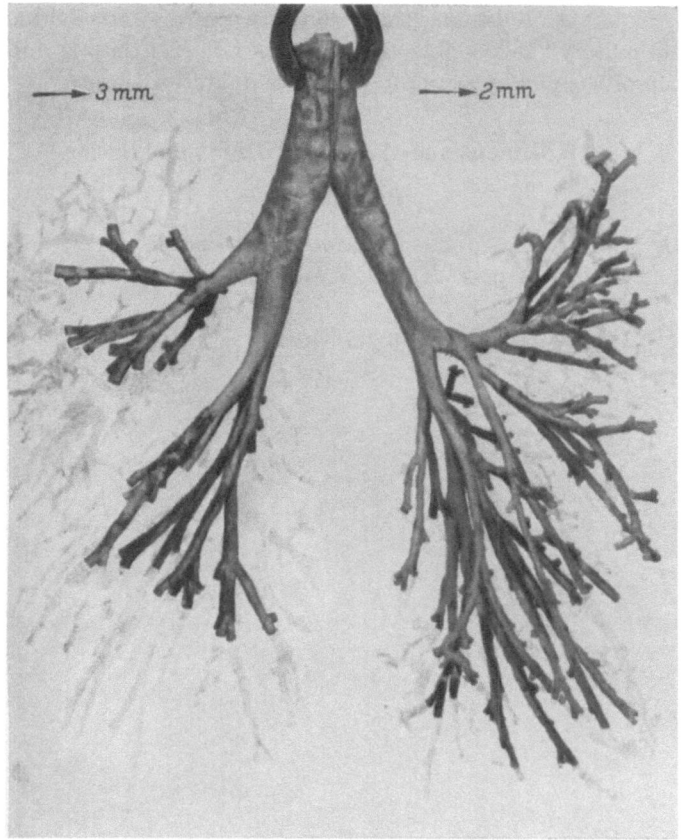

Fig. 99. Anterior view of cast of Fig. 84 terminating with branches of $D^* = 3$ mm and $D^* = 2$ mm

C. Dimensions of Airway Elements of Transitory Zone

Respiratory bronchioles and alveolar ducts and sacs are intimately connected to the large number of alveoli, together with which they form the lung parenchyma, as defined on p. 52. These elements of the transitory zone occur in great number, as will be shown. The unit volume of parenchyma will contain about the same number of elements in any region of the lung, and a random section of parenchyma will cut them in any direction. With respect to the unit volume of parenchyma, the elements of the transitory zone satisfy the condition of randomness of distribution (cf. p. 10), so that random sampling procedures can be used to estimate their number and dimensions.

In histologic sections a clear structural differentiation between alveolar ducts and sacs is not possible. They were, therefore, treated as one group. Peripheral respiratory bronchioles also are not always clearly distinguishable from alveolar ducts; therefore, a clear separation of these classes is likewise not possible. But the number of such uncertain branches is limited to only about 10% of the aggregate ducts and sacs, so that it does not introduce a significant error. Moreover, the degree of "alveolation" of peripheral airways increases gradually toward the periphery, so that a continuous transition between respiratory bronchioles and alveolar ducts exists. The possible error is thus not real, but is due only to a schematization of the airway structures which is unduly rigorous.

1. Dimensions of Alveolar Ducts and Sacs

a) Diameter

The diameter of alveolar ducts and sacs was measured in the five lungs used for microscopic study (see p. 40, Table IV.1). From the set of slides prepared for

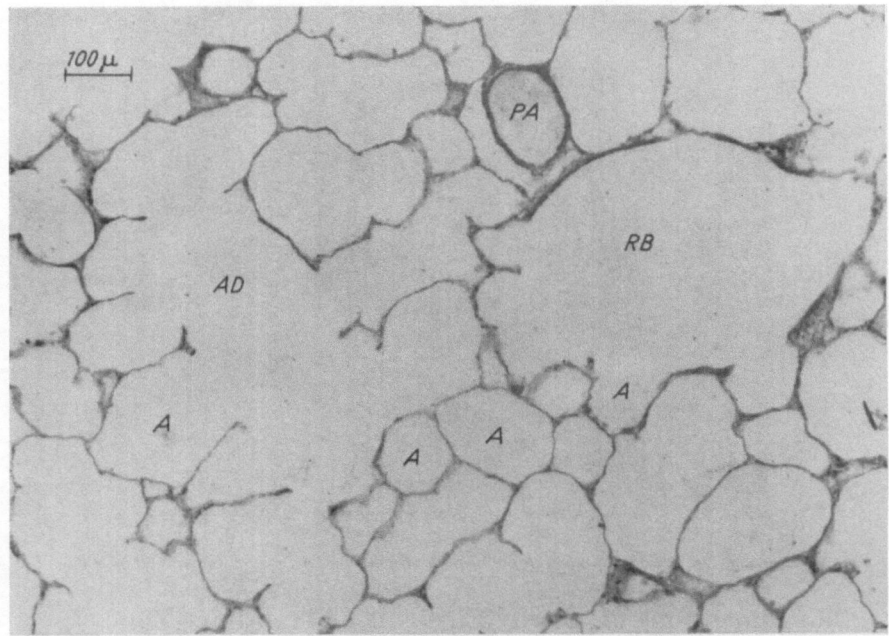

Fig. 100. Histologic section of parenchyma of lung of young woman showing numerous alveoli (*A*) which are partly in relation to alveolar duct (*AD*) or respiratory bronchiole (*RB*). *PA* = branch of pulmonary artery. ×90

each case, five slides were drawn at random, as outlined on p. 45. On each slide at least 20 alveolar ducts or sacs were measured, yielding a total of at least 100 for each case.

In order to make a reliable estimate of the diameter, sections of alveolar ducts which allowed a clear identification of the duct outline and of the direction of cutting were selected for measurement. One such example is illustrated in Fig. 100. On this oblique section the diameter was measured in the center, perpendicular to the long axis. The results of these measurements are illustrated in Fig. 101 and in Table X.2. The diameter of alveolar ducts and sacs varies between 150 and

400 μ in children and from 200 to 600 μ in adults. The standard deviations were 15 to 20% of the mean. In Fig. 102 the average diameters are plotted against the size of the lung measured in terms of lung volume V_L. The relationship between the average size of ducts and the size of lung is quite evident, although it was not possible to develop a relevant formula.

b) Length

No method of estimating directly the length of alveolar ducts and sacs has yet been found. Two-dimensional samples, like sections, are not suitable for this purpose. A seemingly short branch may be partially represented, and an apparently

Table X.2. *Diameter of Transitory Airways*

#	V_L ml	Alveolar ducts and sacs		Respiratory bronchioles	
		$\overline{D}_{ADS}$ 10^{-4} cm	S. D.	$\overline{D}_{RB}$ 10^{-4} cm	S. D.
2471	2580	254	± 44	404	± 51
555	2680	270	± 42	426	± 71
9975	5750	406	± 78	570	± 139
8710	7100	468	± 79	(690	$\pm 280)*$
3361	5950	375	± 58	(810	$\pm 380)*$

* Small number of branches measured.

long branch may be composed of two or even more successive branches, if the intermediate plane of branching would be somewhat perpendicular to the plane of section: in such a case the conjugate branch would lie above or below the section. The usually hazy outline of the central air ducts due to alveolation makes it difficult, if not impossible, to judge such a situation. On the other hand, casting methods

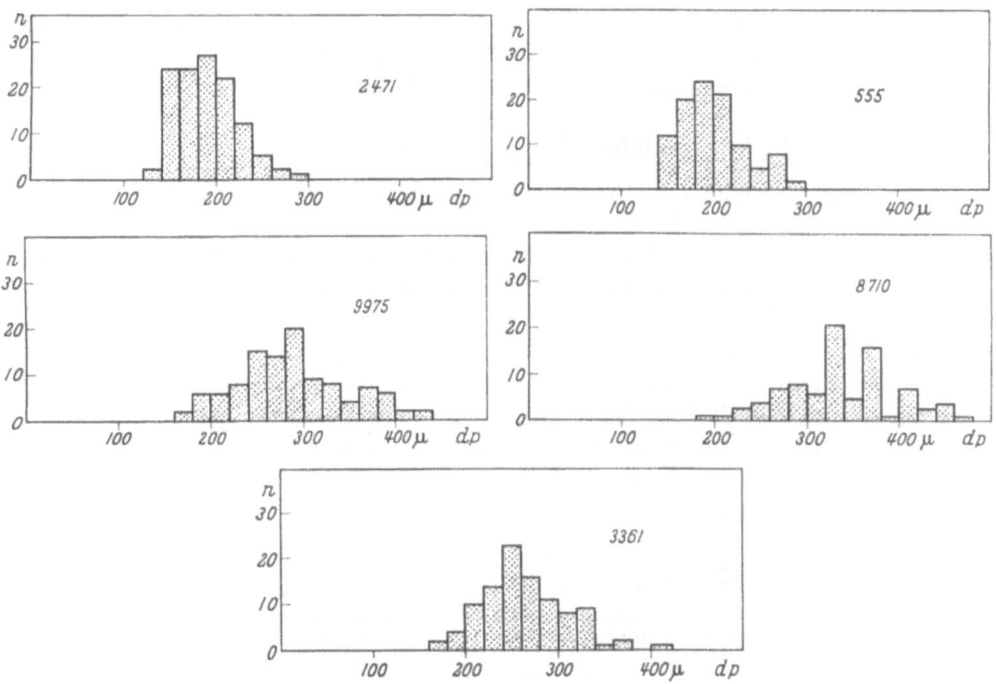

Fig. 101. Distribution of diameter of sections of alveolar ducts and sacs in five cases. Dimensions refer to processed lung

are not suitable for replication of these peripheral airways. The only method would consist of reconstruction from serial sections as done by WILLSON (1922). How-ever, this has not been attempt-ed in the present study.

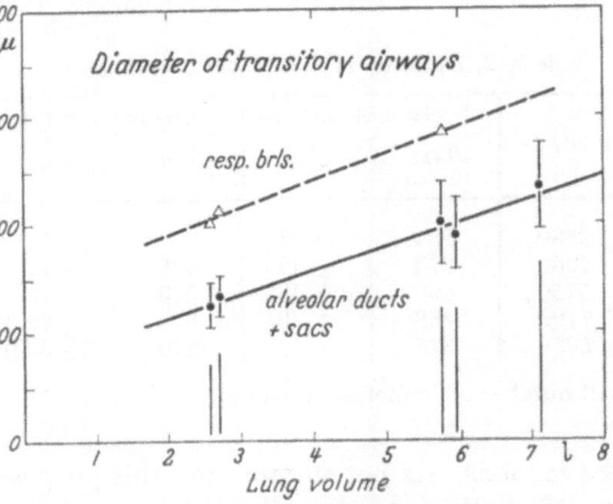

In order to estimate the length, we examined stereo-microscopically the cut surface of thick slices of the fixed lungs used for microscopic investiga-tion. Longitudinal sections of alveolar ducts were studied for their approximate length-to-dia-meter ratio which was found to be 2 on the average, varying from 1 to 3. The alveolar ducts appeared, therefore, to be rela-tively shorter or wider than average bronchi.

Fig. 102. Diameter of transitory airways plotted against lung volume. Average dimensions are derived from distributions of Fig. 101 and refer to fresh lung. Bars indicate standard d eviations

In applying this length-to-diameter ratio to the diameters presented, we found that the average length of alveolar ducts and sacs is roughly 0.7 to 1 mm.

2. Number of Alveolar Ducts and Sacs

Knowing approximately the diameter-to-length ratio of alveolar ducts and sacs, and having determined their volumetric density ϱ_D (Fig. 34), we can use the counting principle outlined on p. 25 to obtain the total number of alveolar ducts and sacs in the entire lung. Equation (3.45) involves three variables which are to be determined:

n_D: *The number of transsections of alveolar ducts and sacs on the unit area of a histologic section.* This was obtained by counting such transsections by a procedure similar to that described for the counting of alveoli. The size of the counting field measured $5.9 \cdot 10^{-2}$ cm²; it included about 20 to 40 duct transsections. On each of five slides randomly selected from each case five fields were counted and the average number per 1 cm² of section was computed (Table X.3).

Table X.3. *Number of Alveolar Ducts and Sacs*

#	Number of transsections per 1 cm² section			Number of ducts per 1 cm³ processed tissue N_D	Total number of ducts and sacs $N_{(D+S)T}$	Average number of airway generations T
	n_D	S. D.	95% confidence interval (% of mean)			
2471	630	±88	±5.4	$14.4 \cdot 10^3$	$13.6 \cdot 10^6$	22.8
555	654	±83	±4.9	$16.1 \cdot 10^3$	$15.8 \cdot 10^6$	22.9
9975	368	±69	±7.2	$6.8 \cdot 10^3$	$14.3 \cdot 10^6$	22.9
8710	336	±47	±5.4	$5.2 \cdot 10^3$	$13.5 \cdot 10^6$	22.8
3361	340	±33	±3.9	$5.3 \cdot 10^3$	$11.6 \cdot 10^6$	22.6

$$\overline{N_{(D+S)T}} = 13.8 \cdot 10^6 \pm 1.5 \cdot 10^6.$$

ϱ_D: *The fraction of the lung volume occupied by ducts* as estimated in Chapter V. It can be read on Fig. 34 from the right-hand scale referring to the parenchyma of the lung, since major airways or blood vessels were excluded from the counting fields.

β_D: *The value of the shape coefficient* β_D which follows from the diameter-to-length ratio $\varepsilon = \dfrac{D}{L}$ (Fig. 17). The value of ε for alveolar ducts was shown to vary between 0.3 and 1 with the highest frequency around 0.5. As already stated, a determination of the actual distribution of ε has not been possible. It was estimated however that approximately 60% of the branches have an ε of about 0.5, while the value is 0.3 or 1.0, respectively, in each half of the remaining 40%. On this basis we estimated a coefficient $1/\beta_D \approx 0.5$ to be appropriate for the present purposes. It was not possible to determine the value of the distribution coefficient K [equation (3.45)] so that we must expect the estimated numbers to be somewhat too small (cf. p. 27).

The number of alveolar ducts and sacs calculated for each case from these data is tabulated in the last column of Table X.3. We observe that it is essentially constant in all five lungs examined, with a group average of

$$\overline{N_{(D+S)T}} = (13.8 \pm 1.5) \, 10^6 \, .$$

From this number and from the number of alveoli estimated above, it follows that approximately 21 alveoli are found per alveolar duct, which is in close agreement with observations made by WILLSON (1922) on serial reconstructions.

Discussion of Possible Errors

The main uncertainty in this determination was the diameter-to-length ratio ε which was approximated in a crude fashion at 0.5 on the average. Assume, now, that the average ε was rather of the order of 1.0 or of 0.3, respectively, allowing an error of $\pm 100\%$. This would give corresponding coefficients $1/\beta_D$ of 0.62 or 0.42. It would not influence the other quantities. In this case the total number of alveolar ducts and sacs would have been estimated at $17.0 \cdot 10^6$ or $11.4 \cdot 10^6$ respectively, a deviation from our figure of $\pm 20\%$. However, we are certain that the error introduced in estimating ε in such crude fashion was not 100%, which indicates that the error affecting the number is even smaller and is within the range of error of the method, which is ± 10 to 20 %. An underestimate by 10 to 20% may have resulted from disregarding the distribution coefficient K (cf. p.27).

We shall use the number of alveolar ducts mainly to deduce the number of generations of airway branching. Since each generation increases the number of branches by a factor of 2, and since alveolar ducts and sacs are the most peripheral branches, even the possible error of 20% affecting the number of ducts will not influence this estimation to any appreciable degree.

3. Dimensions of Respiratory Bronchioles

The number of respiratory bronchioles in the unit volume of parenchyma is evidently much smaller than that of alveolar ducts and sacs (cf. p. 114). It is therefore much more difficult to obtain a significant estimate of their dimensions. In order to obtain an estimate of their diameter all respiratory bronchioles, which were sectioned by the samples used for evaluating alveolar duct dimensions, were measured. The diameters varied considerably; their average values are listed in Table X.2 along with the standard deviations. In the adult they were found to be 600 to 700 μ.

D. Comparison of these Findings with Dimensions Obtained on Bronchograms

A plastic cast is obviously an artificial replica of a human bronchial tree, and it remains to be determined whether its dimensions compare satisfactorily with those obtained on bronchograms of the living. Figure 103 shows one of the broncho-

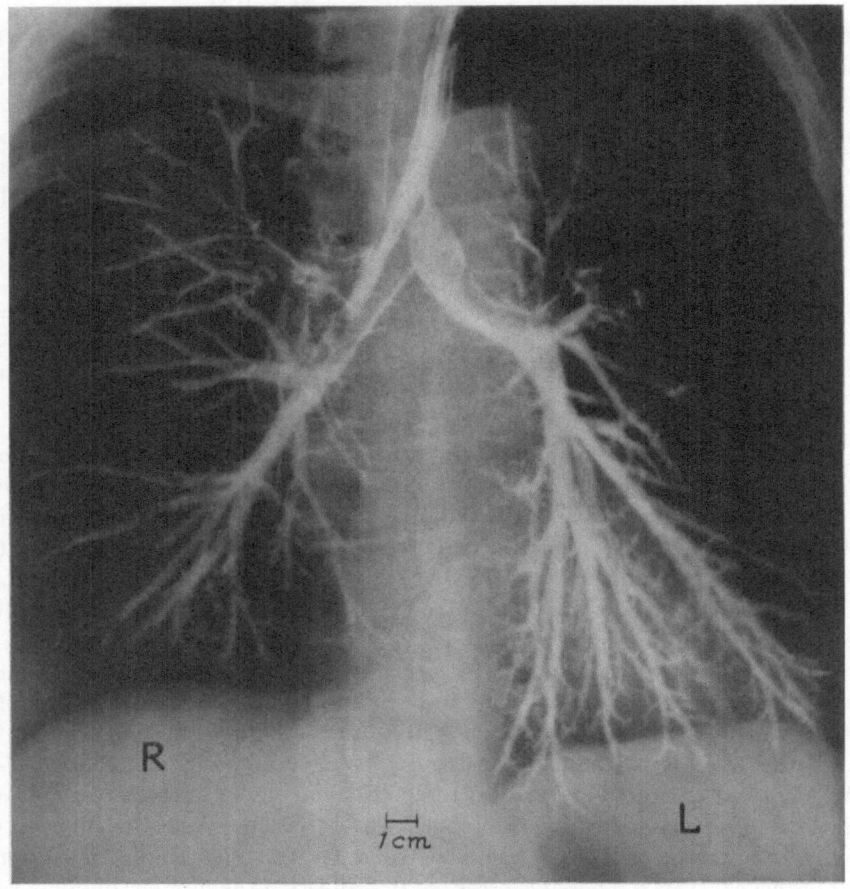

Fig. 103. Bronchogram of young adult used for comparison and for calibration of dimensions obtained on cast. Courtesy Dr. THANE ASCH, New York Hospital and Cornell University. 1:2.5

grams investigated for this purpose. They were prepared by Dr. THANE ASCH of New York Hospital and kindly loaned to us for measurement. Comparison of Fig. 103 with Fig. 84 indicates that the relative dimensions were rather well preserved in our cast. Table X.4 lists the average dimensions of the bronchi in generations 0 to 4, as they could be roughly estimated from this bronchogram.

It is evident that the diameters agree with those of the cast. The lengths check in order of magnitude; it should be noted, however, that the measurement of "length" in a bronchogram is rather uncertain because of the inclination of the branches to the plane of view, and because the origin of branches can rarely be localized with certainty. It is also rather difficult to separate generations. By measuring all visible branches and correcting for an average inclination of smaller branches toward the plane of view, it was possible to estimate the order of magnitude of the airway volume from the bifurcation to bronchi of about 2 mm diameter. This was found to be 30 to 35 ml. On the average, branches of about 2 mm diameter were found in the eighth generation. The calculated volume of our cast from the bifurcation to the eighth generation was 35 ml (Table XI.1), and therefore checks well with the airway volume of this bronchogram. From the chest dimensions we can estimate the total lung volume V_L of this case to be 4500 to 5000 ml. The bronchogram was taken with the patient in a state of moderately deep inspiration.

Table X.4. *Dimensions of Bronchi of First Four Generations as Measured on the Bronchogram of Fig. 103*

Generation z	Average diameter[1] cm	Average length[1] cm
0	2.0	—
1	1.25	2.9
2	0.91	2.0
3	0.6	1.7
4	~0.4	~1.5

Estimated volume of lung: 4.5—5 liter. Estimated total volume of airways from generation 1—4: 30—35 ml.

[1] After correction for enlargement caused by radiography.

These various comparisons indicate that the airway dimensions obtained on our cast compared well with those of an average-sized adult human lung at the standard degree of 3/4 maximal inflation at which the lungs were prepared for the histologic studies (cf. p. 44). We are, therefore justified, in comparing the dimensions for major airways obtained on the cast with those of peripheral structures estimated microscopically on the adult lungs (see p. 134).

E. Average Number of Airway Generations and Correlation of Dimensions

Since airways branch by dichotomy (cf. p. 111) the number of branches in a generation z is expressed by

$$n(z) = 2^z \tag{10.3}$$

as long as all preceding generations were complete, that is, as long as every branch in generations $0 \rightarrow (z-1)$ divided into two branches. On the other hand, we found the total number of alveolar ducts and sacs to be $N_{DST} = 13.8 \cdot 10^6$ on the average (see p. 131, Table X.3). This allows us to estimate the order of generation T at which terminal airways (alveolar sacs) are, on the average, found.

Alveolar sacs are generally preceded by about three generations of alveolar ducts (Fig. 82). The total number of ducts and sacs, therefore, corresponds to the sum of the numbers $n(z)$ in the generations $(T-3)$ to T, and we find

$$N_{DS} = \sum_{z=(T-3)}^{T} 2^z = 15 \cdot 2^{T-3} = \frac{15}{8} \cdot 2^T . \tag{10.17}$$

By solving for T and introducing the numerical value of N_{DS} we obtain

$$T = \frac{\log\left(\frac{8}{15}N_{DS}\right)}{\log 2} = 22.8 . \tag{10.18}$$

Thus, we find that the average number of airway generations is 23, about 20% of the branches being "missing" in the last generation. It should be noted that even gross errors in the determination of the number of alveolar ducts and sacs would not have influenced this finding to an appreciable extent; it would merely have changed the percentage of the branches missing in the 23rd generation. As discussed on p. 131, the maximal possible error was estimated to be of the order of $\pm 20\%$, while an error of at least 50% would be required to change the number of generations significantly. It should also be noted that an error in the assumption that alveolar ducts are distributed over three generations would have little effect. If they would constituted four generations this would add about $0.5 \cdot 10^6$ to the total number, an inappreciable error of 3%. If, as an extreme case, all airways from the trachea to the periphery had been alveolar ducts and had been included in N_{DS}, the order T would have been estimated at 22.6, so that this evaluation is reasonably insensitive to various possible errors.

Due to the irregularity of dichotomy, terminal airways are actually distributed over an entire range of subsequent generations. At present, no reliable method, which would estimate this distribution, is available.

The foregoing estimation of the average number of airway generations indicates the average location of alveolar ducts and sacs within the airway system. It has also been shown by comparison with bronchograms that the average dimensions of conductive airways determined on our cast compared well with those to be expected in an average adult human lung, so that they can be compared with the dimensions of transitory airways determined microscopically (Table X.2). The range of diameters and lengths of alveolar ducts and sacs, and the range of diameters of respiratory bronchioles have therefore been plotted as boxes on Figs. 93 and 94 in relation to generations 20 to 23 and 17 to 19, respectively.

It was observed on p. 123 that equation (10.13) accurately described the decrease of the average diameter of conductive airways in generations 2 to 10. Figure 93 reveals, however, that the diameter of alveolar ducts and sacs deviates greatly from the values expected by extrapolation of the exponential function (10.13). It was not possible to find a physical function that would describe the observed decrease of airway diameters with progressing generations. Empiric functions of purely descriptive character were therefore fitted to the data (Fig. 94). The diameter of major distributing airways of generation 0 to 3 were described by

$$d_1(z) = d_0 \cdot e^{-0.388z} \tag{10.19}$$

where d_0 is the diameter of the trachea. Diameters of intrapulmonary conductive and transitory airways satisfied the equation

$$d_2(z) = d_0' \cdot e^{-(0.293-0.0062\,z)\,z} \tag{10.20}$$

where $d_0' = 1.3$ cm is the intercept at the origin. These two regions correspond to those distinguished on p. 125 for the fitting of two exponential functions to the average airway lengths (Fig. 94b). It can also be seen in Fig. 94b that the length of alveolar ducts is satisfactorily described by equation (10.16). By interpolation these equations indicate the approximate average dimensions of airways in intermediate generations which are not yet amenable to direct measurement.

F. Discussion of the Findings

The dimensions of the major conductive airways were estimated on a plastic cast of a human bronchial tree, and checked against measurements on bronchograms. This permitted the recognition of some regular and irregular features of the airway system. The dimensions of the most peripheral airway elements of the transitory zone were determined on histologic preparations. By interpolation the average dimensions of airways in intermediate generations could be approximated. These findings will now be discussed as they relate to previously available information.

The most complete study of the dimensions of the human airway system was carried out by ROHRER (1915) in the course of his classic work on airway resistance. He determined diameter and length of the bronchial tree of a fresh cadaver lung with the aid of calibrated bougies, which he inserted peripherally while dissecting proximal branches. This method was satisfactory in measuring larger branches, but it was probably not sufficiently reliable toward the periphery. ROHRER lists his data in terms of the number of branches of a given diameter. Comparing them with our findings we observe excellent agreement down to branches of 4 mm in diameter, but it appears that ROHRER greatly underestimated the number of smaller branches: while he found 86 branches of 2 mm, we counted 300 to 400 in this class. This serious underestimate has greatly influenced his further findings on the number of peripheral airways. Thus he arrived at the conclusion that the total number of airways in the entire lung, from main bronchi to alveolar ducts, was approximately one million, which is evidently too low an estimate (cf. p. 139).

ROHRER estimated the dimensions of intralobular airways on the basis of the information given by LAGUESSE and D'HARDIVILLIER (1902) and v. EBNER (1899). He assumed an exponential decrease of the diameters as expressed by equation (10.12); it is argued on p. 134 that this assumption is not justified for the peripheral airways.

POLICARD and GALY (1945) indicated dimensions of the major airways which compared well with our findings. REID (1958) analyzed the relative dimensions of peripheral airways in bronchograms. Although no detailed comparison of the findings is possible, they appear to show no gross disagreement. STUTZ (1949) measured the bronchial shadows of bronchograms prepared during deep inspiration and expiration. Generally speaking, his data coincide with our findings, although a detailed comparison is difficult.

The irregularity of airway branching was carefully investigated by Ross (1957) on the dog's lung. The distribution pattern of "terminal bronchi" of equal diameter with respect to the bronchial pathway length appears quite similar to that presented in Fig. 97 for the distance Δ of airways of equal diameter D^* from the trachea. A more detailed comparison is barely possible, since Ross's data refer to the dog's lung, whose bronchial tree has an architecture different from that of the human airway system.

Chapter XI

Geometric and Dimensional Airway Models of Conductive, Transitory and Respiratory Zones of the Human Lung

The rational analysis of the physical events associated with the movement of gases in the lung makes use of relatively refined models which reflect some of the properties of the airway system. Some investigators used simple airway models which did not need to be dimensionally defined (OTIS et al., 1956; FRY and HYATT, 1960; and others). ROHER (1915) measured diameter and length of the elements of the bronchial tree (see p. 135) and constructed a dimensional model on which he based his reasoning on flow resistance in the human airways. In his study on the deposition of airborne particles in the lung, FINDEISEN (1935) proposed and used a dimensional airway model which has been widely applied subsequently (LAN-DAHL, 1959; and others) since it appeared to be the most complete information available (cf. DITTMER and GREBE, 1958). Unfortunately, Findeisens' model conveys much erroneous information. It was not based on sufficient actual data and thus the dimensions of even the most proximal airways are not correct. In addition, the grouping of subsequent airways into nine categories with different "branching factors" is entirely inadequate to represent the pulmonary architecture. In studying the distribution of ventilation in the dog's lung ROSS (1957) measured the dimensions of a bronchial tree and designed a model which took special account of the irregular features of the dog's airway system. This model is not directly applicable to the human lung because of significant architectural and dimensional differences in the two species.

It appears desirable therefore (MEAD, 1961; GOMEZ, 1962), to propose new models of the human airway system on the basis of the foregoing analysis. It seems appropriate to design two different models, the first emphasizing the *regular* features of the airways, the second accounting for some of their *irregularities*. Both models will be based on the fundamental architectural pattern of dichotomy, as decribed on p. 111. They will refer to an average adult human lung, and will comprise the airways of the conductive, transitory, and respiratory zone, as well as the blood vessels (alveolar capillaries) of the respiratory zone.

A. Model "A": Regular Dichotomy

For many considerations of the over-all behavior of the airway system it may be adequate to disregard its irregularities and to concentrate on its regular features. We can therefore construct our first lung model (Fig. 104) by assuming that the airways multiply by regular dichotomy (see p. 111) so that all elements (branches) in a given generation z have identical dimensions. In addition, all generations are complete. The airways of this model lung thus branch over 23 generations (see p. 134). As indicated on p. 9 and in Figs. 6 and 82, these generations can be grouped into zones with respect to the structural properties of their elements. The conductive zone will include generations 0 to 16, while generations 17 to 19 (respiratory bronchioles) and 20 to 23 (alveolar ducts and sacs) will belong to the transitory zone and bear alveoli.

The number of elements in each generation z is evidently

$$n(z) = 2^z .$$ (11.1)

Diameter and length of the single cylindric elements are given by the pairs of empiric equations fitted to the measurements obtained on cast and histologic

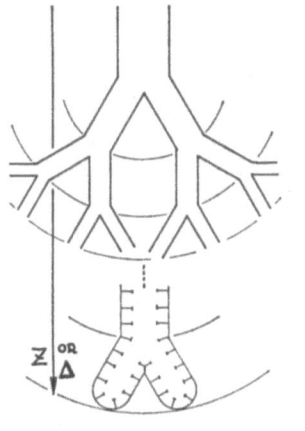

Lung model A	
$z \leqq 3$	$z > 3$
$d_1(z) = d_0 \cdot e^{-a \cdot z}$	$d_2(z) = d_0' \cdot e^{-(b-cz)z}$
$l_1(z) = l_0 \cdot e^{-m \cdot z}$	$l_2(z) = l_0' \cdot e^{-u \cdot z}$
$S_1(z) = S_0 \cdot e^{-A \cdot z}$	$S_2(z) = S_0' \cdot e^{(B+C \cdot z)z}$
$V_1(z) = V_0 \cdot e^{-E \cdot z}$	$V_2(z) = V_0' \cdot e^{(C \cdot z - F) \cdot z}$

Lung volume (air): 4800 ml
Respiratory zone: 3150
Transitory zone: 1500
Conductive zone: 150

23 generations regular dichotomy

Fig. 104. Diagram of arrangement of elements in lung model "A" with set of equations defining its dimensions. Compare text (Note: S signifies cross sectional area in this diagram)

preparations (p. 125, and p. 134, Fig. 94). Table XI.1 lists these equations along with the formulas (11.6) to (11.9) for the over-all cross section $A(z)$ and the over-all volume $V_{(z)}$ of the airways in each generation as they could be derived

Table XI.1. *Airway Model Equations*

$$D(z) = \begin{cases} D_0 \cdot e^{-0.388 z} & \text{if } z \leqq 3 \quad D_0 = 1.8 \text{ cm} \quad (11.2) \\ D_0' \cdot e^{-(0.2929 - 0.00624 z) z} & \text{if } z > 3 \quad D_0' = 1.3 \text{ cm} \quad (11.3) \end{cases}$$

$$L(z) = \begin{cases} L_0 \cdot e^{-0.92 z} & \text{if } z \leqq 3 \quad L_0 = 12 \text{ cm} \quad (11.4) \\ L_0' \cdot e^{-0.17 z} & \text{if } z > 3 \quad L_0' = 2.5 \text{ cm} \quad (11.5) \end{cases}$$

$$A(z) = \begin{cases} A_0 \cdot e^{-0.083 z} & \text{if } z \leqq 3 \quad A_0 = 2.54 \text{ cm}^2 \quad (11.6) \\ A_0' \cdot e^{(0.1074 + 0.0125 z) z} & \text{if } z > 3 \quad A_0' = 1.32 \text{ cm}^2 \quad (11.7) \end{cases}$$

$$V(z) = \begin{cases} V_0 \cdot e^{-1.003 z} & \text{if } z \leqq 3 \quad V_0 = 30.5 \text{ cm}^3 \quad (11.8) \\ V_0' \cdot e^{(0.0125 z - 0.0626) z} & \text{if } z > 3 \quad V_0' = 3.3 \text{ cm}^3 \quad (11.9) \end{cases}$$

$$\sum_{i=0}^{z} V_i = \sum_{i=0}^{3} V_0 \cdot e^{-1.003 i} + \sum_{i=4}^{z} V_0' \cdot e^{(0.0125 i - 0.0626) i}$$ (11.10)

from equations (11.2) to (11.5). Equation (11.10) expresses the accumulated volume of the airways from the larynx to generation z. These functions have been plotted in Figs. 105 and 106 respectively.

This airway model refers to an average-sized adult human lung at a degree of 3/4 maximal inflation (cf. p. 44). It has a total air volume of about 4,800 ml of which 66% is contained in the alveoli (see below). Table XI.2 lists the numerical values of the airway dimensions of this lung. It should be noted that the conductive zone has a total volume of about 150 ml, while the transitory zone has a volume of 1,500 ml.

The 300 million alveoli of the respiratory zone are laterally apposed to the elements of the transitory zone (Figs. 37, 41, 104). The surface of alveolar ducts and sacs is completely alveolated, that of respiratory bronchioles (generations 17 to 19) only partially. The fraction $\eta(z)$ of the surface of each element which opens into alveoli is listed in Table XI.3, together with the number $n_{A/D}$ of alveoli associated with each element and the number $n_A(z)$ of alveoli per generation z. By assigning to each model alveolus the dimensions given on p. 109 and in Table IX.1, we can define some over-all respiratory parameters of the total lung model, for example, the air-tissue-interface $I_{A-T}(z)$, the tissue-blood-interface $I_{T-B}(z)$ and the capillary blood volume $V_c(z)$ in each generation. All these values are listed in Table XI.3.

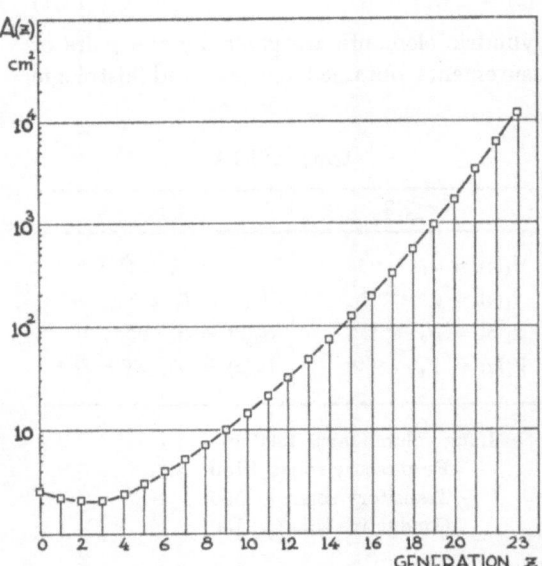

Fig. 105. Total airway cross section in each generation z

In the model of the respiratory zone the single alveolus was geometrically described as a fraction of a sphere (p. 105). For the present model the alveoli are regarded as wedges of the alveolar sleeve enveloping the alveolar ducts (Fig. 41a and 107). The respiratory surface is contributed by the cylindric surface of the alveolar "dome" and by the two sides of the lateral septa (Fig. 107b). These septa are placed in such a fashion that any cylindric slice of height $d\Delta$ (Fig. 107a) of an alveolar duct contains the same elemental area of the respiratory surface

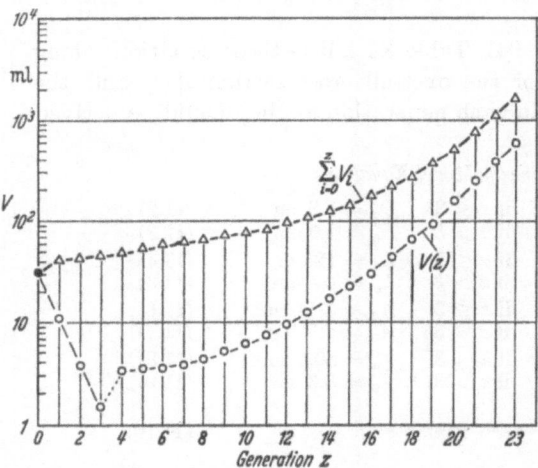

Fig. 106. Total airway volume $V(z)$ in each generation and cumulated airway volume $\sum_{i=0}^{z} V_i$ from the origin of the trachea to generation z inclusively

$$S_E = 0.625 \cdot \eta(z)\, d\Delta .\tag{11.11}$$

The sum of all such elemental slices, which are found at equidistant points in any generation z, have a total respiratory surface area

$$S_z(d\Delta) = 0.625 \cdot \eta(z) \cdot d\Delta \cdot e^{0.693 z} .\tag{11.12}$$

Table XI.2. *Dimensions of Human Airway Model "A"*
Average adult lung with volume 4800 ml at about 3/4 maximal inflation

Gener-ation	Number per generation	Diameter	Length	Total crossection	Total volume	Accumul. volume
z	n(z)	d(z)	l(z)	S(z)	V(z)	$\sum_{i=0}^{z} V(i)$
		cm	cm	cm²	cm³	cm³
0	1	1.8	12.0	2.54	30.50	30.5
1	2	1.22	4.76	2.33	11.25	41.8
2	4	0.83	1.90	2.13	3.97	45.8
3	8	0.56	0.76	2.00	1.52	47.2
4	16	0.45	1.27	2.48	3.46	50.7
5	32	0.35	1.07	3.11	3.30	54.0
6	64	0.28	0.90	3.96	3.53	57.5
7	128	0.23	0.76	5.10	3.85	61.4
8	256	0,186	0.64	6.95	4.45	65.8
9	512	0.154	0.54	9.56	5.17	71.0
10	1024	0.130	0.46	13.4	6.21	77.2
11	2048	0.109	0.39	19.6	7.56	84.8
12	4096	0.095	0.33	28.8	9.82	94.6
13	8192	0.082	0.27	44.5	12.45	106.0
14	16384	0.074	0.23	69.4	16.40	123.4
15	32768	0.066	0.20	113.0	21.70	145.1
16	65536	0.060	0.165	180.0	29.70	174.8
17	131072	0.054	0.141	300.0	41.80	216.6
18	262144	0.050	0.117	534.0	61.10	277.7
19	524288	0.047	0.099	944.0	93.20	370.9
20	1048576	0.045	0.083	1600.0	139.50	510.4
21	2097152	0.043	0.070	3220.0	224.30	734.7
22	4194304	0.041	0.059	5880.0	350.00	1084.7
23*	8388608	0.041	0.050*	11800.0	591.00	1675.0

* Adjusted for complete generation.

Table XI.3. *Dimensions of Respiratory Zone of Human Airway Model "A"*

Gener-ation z	Fraction of surface alveolated η(z)	Number of alveoli per		Air-tissue interface $I_{A-T}(z)$ cm²	Capillary volume $V_C(z)$ cm³	Tissue-blood interface $I_{T-B}(z)$ cm²
		duct	generation			
0	0.0					
...						
16	0.0					
17	0.12	5	0.6 · 10⁶	0.16 · 10⁴	0.28	0.14 · 10⁴
18	0.25	8	2.0 · 10⁶	0.54 · 10⁴	0.97	0.47 · 10⁴
19	0.50	12	6.0 · 10⁶	1.62 · 10⁴	2.90	1.40 · 10⁴
20	1.00	20	21.0 · 10⁶	5.70 · 10⁴	9.80	4.90 · 10⁴
21	1.00	20	41.5 · 10⁶	11.30 · 10⁴	19.60	9.70 · 10⁴
22	1.00	20	84.0 · 10⁶	22.70 · 10⁴	39.20	19.60 · 10⁴
23*	1.00	17*	143.0 · 10⁶	38.60 · 10⁴	67.20	33.40 · 10⁴

* Adjusted for complete 23rd generation.

The corresponding individual and total volumes of the capillary bed in these airway elements are

$$v_{CE} = 1.17 \cdot 10^{-4} \cdot \eta(z) \cdot d\Delta , \tag{11.13}$$

$$V_{cz}(d\Delta) = 1.17 \cdot 10^{-4} \eta(z) \, d\Delta \cdot e^{0.693z} . \tag{11.14}$$

Other parameters of respiratory significance can be derived on the basis of the relations given in Chapters VI, VII and IX. Since $\eta(z) = 0$ for $0 \leqq z < 17$, $S_z(d\varDelta)$ will be zero for all generations from 0 to 16. It should be noted that z is a discrete variable which can only assume integer values from 0 to 23.

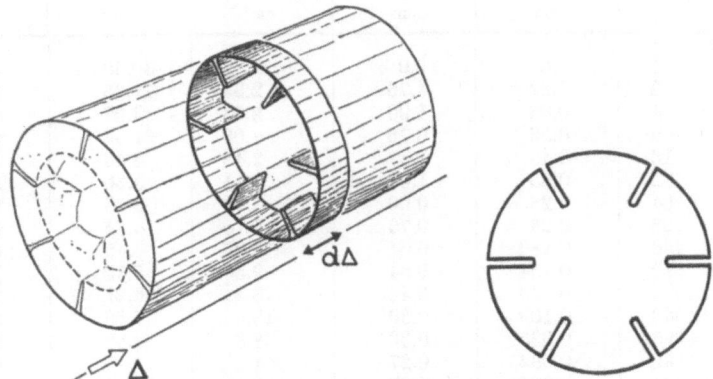

Fig. 107. Arrangement of respiratory elements in lung model "A". Compare text

B. Model "B": Irregular Dichotomy

Suppose the lung to be subdivided into n_u units, each one being supplied by an entrance bronchus of equal diameter D^*, and assume that these units have an equivalent volume v_u and each contains an equivalent number of alveoli. If we take as point of departure $D^* = 2$ mm, we find $n_u = 400$, and accordingly $v_u = 12$ ml. Figure 108 shows how this volume is distributed over the three airway zones.

In Fig. 96 it was shown that bronchi of diameter $D^* = 2$ mm were found in generations 4 to 13; their density in each generation z could be well approximated by a binomial distribution:

$$f_2(z) = \frac{10!}{(14-z)!\,(z-4)!} \cdot p^{(z-4)}(1-p)^{(14-z)} \tag{11.15}$$

where $p = 0.46$. As demonstrated in Fig. 97, the number of branches with $D^* = 2$ mm found at a distance $\varDelta$ from the root of the trachea was well represented by a normal distribution:

$$F_2(\varDelta) = 0.157 \cdot e^{-0.078(\varDelta-24.5)^2}. \tag{11.16}$$

The functions $f_2(z)$ and $F_2(\varDelta)$ thus fix the location of the entrance bronchi to these $n_u = 400$ units of equal volume. These will originate in generations 4 to 13 and at varying distances from the root of the trachea, ranging from 18 to 31 cm.

Within each unit the airways branch over about 14 more generations, on the average, until alveolar sacs are reached. In first approximation these can be assumed to follow the pattern of regular dichotomy, so that their dimensions would follow from functions (11.3) and (11.5) for values of $8 < z \leq 23$.

A further approximation of the irregularity of airway branching can be achieved through the following reasoning: (1) The distribution of branches of $D^* = 2$ mm with respect to $\varDelta$ was rather well described by the normal curve

$F_2(\varDelta)$ given in equation (11.16). We can, therefore, postulate that the density of branches with $D* = 1$ mm and $D* = 0.5$ mm also follows a normal function of the general form

$$F(\varDelta) = \frac{1}{\sqrt{2\pi\,\sigma^2}} \cdot e^{-\frac{1}{2}\frac{(\varDelta - \overline{\varDelta})^2}{\sigma^2}} \qquad (11.17)$$

where $\overline{\varDelta}$ is the average distance and σ^2 is the variance. (2) In Fig. 98 the variation of the distance $\varDelta'$ of branches of $D_2^* = 2$ mm from those of $D_4^* = 4$ mm was

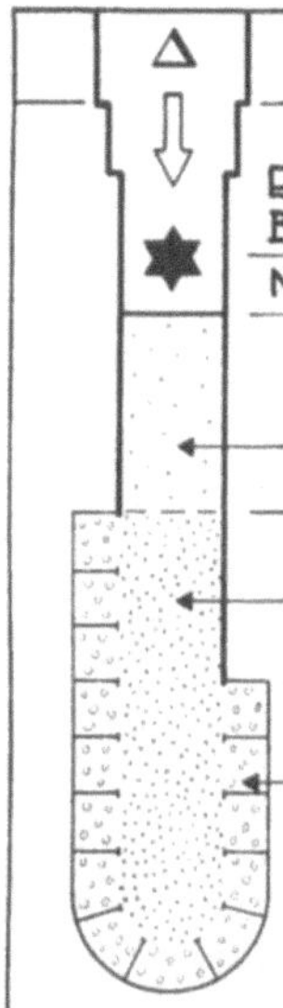

△ ⇩ ✦	LUNG UNITS		
ENTRANCE BRONCHUS $D*$	3 mm	2.5 mm	2 mm
NUMBER UNITS	100	200	400
VOLUMES	ml	ml	ml
CONDUCTIVE ZONE	1	0.35	0.16
TRANSITORY ZONE	15	7.5	3.75
RESPIRATORY ZONE	31	16	8.1
TOTAL VOLUME	47	24	12
NUMBER OF ALVEOLI	$30 \cdot 10^5$	$15 \cdot 10^5$	$7.4 \cdot 10^5$

Fig. 108. Dimensions of lung units of the irregular lung model "B" defined by diameter $D*$ of entrance bronchus. Compare text

shown to be well described by a normal curve $G(\varDelta')$; it should be noted that the error introduced by the fact that the normal function extends into the negative range of $\varDelta'$ is inappreciable. We find that $\overline{\varDelta}' = 4.5$ cm, and $\sigma' = 1.8 = 0.4 \cdot \overline{\varDelta}'$, so that

$$G(\varDelta') = \frac{1}{1.8 \cdot \sqrt{2\pi}} \cdot e^{-\frac{1}{2}\frac{(\varDelta' - 4.5)^2}{3.24}} \qquad (11.18)$$

(3) We next postulate that $G(\varDelta')$ is representative of the distribution of distances $\varDelta'$ from branches with diameter $D*$ to peripheral branches with diameter $1/2\,D*$. From functions (11.3) and (11.5) we find the average airway length $\varLambda_{4/2}$ which is necessary to reduce an average diameter of 4 mm to one of 2 mm as 2.6 cm in a scheme of regular dichotomy. We thus derive that $\overline{\varDelta}'_{4/2} = 1.7 \cdot \varLambda_{4/2}$. The above postulate requires that the general value of $\overline{\varDelta}'$ can be estimated from the corresponding value of $\varLambda$, taken from (11.3) and (11.5), through the coefficient 1.7, and that the general standard deviation is 40% of $\overline{\varDelta}'$. Performing this for the intervals $D*\,(2 \to 1)$ and $D*\,(1 \to 0.5)$ we obtain the following parameters

$$\overline{\varDelta}'_{2/1} = 3.5 \text{ cm}, \qquad \sigma_{2/1} = 1.4 \text{ cm}$$
$$\overline{\varDelta}'_{1/0.5} = 2.25 \text{ cm}, \qquad \sigma_{1/0.5} = 0.9 \text{ cm}$$

which define corresponding distributions $G_{2/1}(\varDelta')$ and $G_{1/0.5}(\varDelta')$. By convolution of these functions with $F_2(\varDelta)$ we derive the frequency distributions $F_1(\varDelta)$ and $F_{0.5}(\varDelta)$ of the distance $\varDelta$ from the root of the trachea to airways of diameter $D^* = 1$ and 0.5 mm. These curves are plotted on Fig. 109.

The values for D^* of 1 mm and 0.5 mm were arbitrarily selected so that the corresponding units can be defined only approximately. We estimate that there are from 2,000 to 2,500 units with an entrance bronchus or bronchiolus of 1 mm;

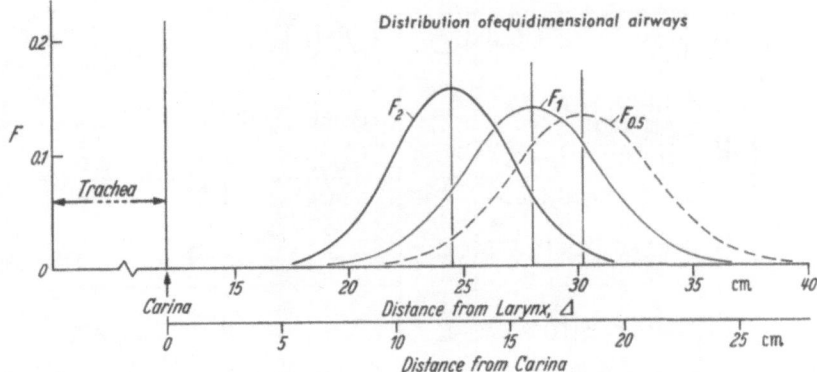

Fig. 109. Distribution of airways of diameter D^* with respect to distance $\varDelta$ from the origin of the trachea. F_2: $D^* = 2$ mm, normal curve derived from data (Fig. 97); F_1: $D^* = 1$ mm, $F_{0.5}$: $D^* = 0.5$ mm, both derived hypothetically. Compare text. Curves suggest distribution of lung units

they would have a volume of about 2 to 2.5 ml, and their airways would branch over an additional 11 generations. Airways of 0.5 mm diameter should correspond to respiratory bronchioles of the first order; their number is estimated to be of the order of 150,000. The volume of these units, which corresponds about to acini (ENGEL, 1958; and others), is about $3 \cdot 10^{-2}$ ml. There are approximately 6 more generations of branching beyond their entrance bronchus. Figure 108 presents diagrammatically the manner in which the volume of these units is distributed over the three airway zones.

Conclusion

In these pages we have analysed the design and dimensions of typical pulmonary structures which appeared to be of primary functional significance in ensuring adequate conditions for gas exchange between air and blood in the lung. These studies have yielded two major results. Firstly, it became apparent that the human lung must be constructed according to a well specified plan: The airway system was found to branch over the same average number of generations, and the number of structural elements, such as alveoli or capillaries, was shown to be essentially identical for all lungs investigated, regardless of the size of the organ. Secondly, it has been possible to formulate fundamental dimensional relations between various structural parameters and the lung size, which led to the establishment of geometrical and dimensional models of the human lung.

It is quite evident that this monograph is not a final report, but rather a sketchy proposition of a line of morphologic investigation which still requires much work before it can be considered fully developed. In particular, too little is known as yet about the changes of pulmonary dimensions with varying functional conditions; and we have only a vague idea about the geometry of lung structures during life. It may be anticipated that some of these problems will be clarified by combining the approach outlined here with methods of "supravital" tissue preparation by ultrarapid freezing as used, for example, by Staub and Storey in a recent study on the lung.

In this presentation no attempt was made to interpret the morphometric findings from a functional point of view. This is left to the biophysicist who may wish to introduce the data documented here into his theoretically derived physical relations together with functional data furnished by experimental physiology. This would establish to what extent the structural design of the respiratory apparatus can influence its performance, and what limitations it may impose, in health and disease, on the exchange of an adequate amount of gas between air and blood.

As morphologists we may also reverse the proposition and ask to what extent functional requirements may determine the quantitative design of the respiratory organ. We see mainly two avenues which could provide an answer to this question: In a comparative study Tenney and Remmers have recently found a direct correlation between the relative alveolar surface area and the relative oxygen need of various mammalian species. This line of work promises many more significant findings if pursued in greater detail on a variety of structural parameters. In addition, the experimental imposition of various chronic conditions on growing and adult laboratory animals and the subsequent morphometric study of their lungs may show to what extent and by what means the respiratory apparatus can adapt itself structurally to altered functional requirements.

Bibliography

Aeby, C.: Der Bronchialbaum der Säugetiere und des Menschen. Leipzig 1880.

Altmann, K.: Experimentell-morphologische Untersuchungen über die Beziehung zwischen Lungenkapillarweite und dem Lungendehnungsgrad. Z. ges. exp. Med. 122, 516—548 (1954).

Arthus, 1927: Quoted from Bargmann 1936.

Baldwin, E. de F., A. Cournand and D. W. Richards jr.: Pulmonary insufficiency: I. Methods of analysis, physiologic classification, standard values in normal subjects. Medicine 27, 243—278 (1948).

Bargmann, W.: Die Lungenalveole. In: Handbuch der mikroskopischen Anatomie 5/3, 799—859, 1936.

—, and A. Knoop: Vergleichende elektronenmikroskopische Untersuchungen der Lungenkapillaren. Z. Zellforsch. 44, 263—281 (1956).

Barnhard, H. J., J. A. Pierce, J. W. Joyce and J. H. Bates: Roentgenographic determination of total lung capacity. A new method evaluated in health, emphysema and congestive heart failure. Amer. J. Med. 28, 51—60 (1960).

Bates, D. V., C. J. Varvis, R. E. Donevan and R. V. Christie: Variations in the pulmonary capillary blood volume and membrane diffusion component in health and disease. J. clin. Invest. 39, 1401—1412 (1960).

Bender, K. W.: Zur synthetischen Morphologie der Lungen, abgeleitet aus ihrer Entwicklungsgeschichte. Verh. dtsch. Ges. Anat. 32, 97—111 (1923).

Bertalanffy, F. D., and C. P. Leblond: Structure of respiratory tissue. Lancet 269, 1365 to 1368 (1955).

BLUMENTHAL, B. J., and H. G. BOREN: Lung structure in three dimensions after inflation and fume fixation. Amer. Rev. Tuberc. **79**, 764—772 (1959).

CHALKLEY, H. W.: Method for the quantitative morphologic analysis of tissues. J. nat. Cancer Inst. **4**, 47 (1943).

— J. CORNFIELD and H. PARK: A method for estimating volume surface ratios. Science **110**, 295 (1949).

CHASE, W. H.: Distribution and fine structure of elastic fibres in mouse lung. Exp. Cell Res. **17**, 121—130 (1959).

— The surface membrane of pulmonary alveolar walls. Exp. Cell Res. **18**, 15—28 (1959).

CLAUS, M.: Über den feineren Gefäßaufbau gesunder und kranker Lungen. Z. mikr.-anat. Forsch. **37**, 245—258 (1935).

CLEMENTS, J.: Air space dimensions, geometry and growth. In: Normal and abnormal respiration in children, Report 37th Ross Conference on Pediatric Research, 1960, p. 39.

COSENTINO, A., W. STOREY and N. C. STAUB: Small vessel hematocrit in the lungs. Fed. Proc. **20**, 426 (1961).

CROFTON, W.: Probability. In: Encyclopedia Britannica, 9th Ed., 1898.

DELESSE, M. A.: Procédé mécanique pour determiner la composition des roches. C. R. Acad. Sci. (Paris) **25**, 544—545 (1847).

DITTMER, D. S., and R. M. GREBE, ed.: Handbook of Respiration. p. 17. Philadelphia: Saunders 1958.

DREISSENS, J., A. DUPONT and A. DEMAILLE: The ultrastructure of the rat lung examined by the electron microscope. C. R. Soc. Biol. (Paris) **153**, 611—613 (1959).

EBNER, V. v.: Von den Lungen. Keollikers Handbuch der Gewebelehre des Menschen. **3**, 299—314 (1902).

ELIAS, H., A. HENNIG and P. M. ELIAS: Some methods for the study of kidney structure. Z. wiss. Mikr. **65**, 70—82 (1961).

ELZE, C., and A. HENNIG: Die inspiratorische Vergrößerung von Volumen und innerer Oberfläche der menschlichen Lunge. Z. Anat. Entwickl.-Gesch. **119**, 457—469 (1956).

ENGEL, S.: The lobulation of the mammalian lung. Anat. Anz. **106**, 86—89 (1959).

EWART, W.: The bronchi and pulmonary blood vessels, their anatomy and nomenclature. With a critism of Prof. AEBY's views on the bronchial tree of mammalia and of man. London: Churchill 1889.

FINDEISEN, W.: The deposition of small air-borne particles in the human lung during respiration. Pflügers Arch. ges. Physiol. **236**, 367 (1935).

FÖRSTER, R. E.: Exchange of gases between alveolar air and pulmonary capillary blood: pulmonary diffusing capacity. Physiol. Rev. **37**, 391 (1957).

— The pulmonary capillary bed: Volume, area and diffusing characteristics. In: Pulmonary Circulation, p. 45—56. New York: Grune and Stratton 1959.

FREY, 1859: Quoted from BARGMANN 1936.

FRY, D. L., and R. E. HYATT: Pulmonary Mechanics. A unified analysis of the relationship between pressure, volume and gasflow in the lungs of normal and diseased human subjects. Amer. J. Med. **39**, 672—689 (1960).

GERSTEL, G.: Über die Möglichkeiten exakter Bestimmung der Relation zwischen Gewebsbestandteilen (Relative Histometrie mit dem Leitzschen Integrationstisch). Verh. dtsch. Ges. Path. **32**, 140—142 (1950).

GERTZ, H.: Über die Größe der Atmungsfläche der Lungen. Z. Biol. **88**, 172 (1928).

GIESE, W.: Alterslunge und Altersemphysem. Medizin **50**, 2447—2454 (1959).

—, and R. GIESEKING: Die submikroskopische Struktur des fibrillären Grundgerüstes der Alveolarwand. Beitr. path. Anat. **117**, 17—31 (1957).

GIESEKING, R.: Elektronenoptische Beobachtungen im Alveolarbereich der Lunge. Beitr. path. Anat. **116**, 177—199 (1956).

GLAGOLEFF, A. A.: On the geometrical methods of quantitative mineralogic analysis of rocks. Trans. Inst. Econ. Min. Moskau **59**, (1933).

GOMEZ, D. M.: Physico-mathematical analysis of respiratory function. (In preparation).

— A mathematical treatment of the distribution of tidal volume throughout the lung. Proc. nat. Acad. Sci. (Wash.) **49**, 312—319, (1963).

DE GROODT, M., A. LAGASSE and M. SEBRUYNS: Fine structure of the alveolar wall of the lung. Nature (Lond.) 181, 1066—1067 (1958).

GRUEN, F., W. STAMM and K. BUCHER: Probleme des Gasaustausches in der Lunge. Helv. physiol. Pharmacol. Acta 14, 141—153 (1956).

HALES, S.: Vegetable Staticks: Analysis of the air. 1731, p. 240.

HARTROFT, W. S., and C. C. MACKLIN: Intrabronchial fixation of the human lung for purposes of alveolar measurements using 25μ microsections made therefrom. Trans. roy. Soc. Can. (Biol.) 37, 75 (1943).

HAUG, H.: Die Treffermethode, ein Verfahren zur quantitativen Analyse im histologischen Schnitt. Z. Anat. Entwickl.-Gesch. 118, 302—312 (1955).

HAYEK, H. v.: Die menschliche Lunge. Berlin-Göttingen-Heidelberg: Springer-Verlag 1953.

HENNIG, A.: Bestimmung der Oberfläche beliebig geformter Körper mit besonderer Anwendung auf Körperhaufen im mikroskopischen Bereich. Mikroskopie 11, 1—20 (1956).

— Diskussion der Fehler bei der Volumenbestimmung mikroskopisch kleiner Körper oder Hohlräume aus den Schnittprojektionen. Z. wiss. Mikr. 63, 67—71 (1956).

— Fehler der Oberflächenbestimmung von Kernen bei endlicher Schnittdicke. Mikroskopie 12, 7—11 (1957).

— Das Problem der Kernmessung. Eine Zusammenfassung und Erweiterung der mikroskopischen Meßtechnik. Mikroskopie 12, 174—202 (1957).

— A critical survey of volume and surface measurements in microscopy. Zeiss Werkzeitschrift No. 30, 1959.

HIERONYMI, G.: Über den durch das Alter bedingten Formwandel menschlicher Lungen. Ergebn. allg. Path. path. Anat. 41, 1—62 (1961).

HOLMES, A. H.: Petrographic methods and calculations. London: Murby & Co. 1927.

HUSCHKE, 1844: Quoted from BARGMANN 1936.

JACKSON, C. L., and J. E. HUBER: Correlated applied anatomy of the bronchial tree and lungs with a system of nomenclature. Dis. Chest 9, 319 (1943).

JUNGHANSS, W.: Die Endstrombahn der Lunge im postmortalen Angiogramm. Virchows Arch. path. Anat. 331, 263—275 (1958).

KARNOVSKY, M. J.: Simple methods for "staining with lead" at high pH in electronmicroscopy. J. biophys. biochem. Cytol. 11, 729—732 (1961).

KARRER, H. E.: An electronmicroscopic study of the fine structure of pulmonary capillaries and alveoli of the mouse. Bull Johns Hopk. Hosp. 98, 65—83 (1956).

— Ultrastructure of mouse lung; fine structure of capillary endothelium. Exp. Cell Res. 11, 542—547 (1956).

— Ultrastructure of mouse lung; mitochondria and endoplasmic reticulum of bronchiolar cells. J. biophys. biochem. Cytol. 2, 115—118 (1956) (Suppl.).

— Ultrastructure of mouse lung; alveolar basement membrane. J. biophys. biochem. Cytol. 2, 287—292 (1956) (Suppl.).

— Ultrastructure of mouse lung; general architecture of capillary and alveolar walls. J. biophys. biochem. Cytol. 2, 241—252 (1956).

— The ultrastructure of mouse lung: The alveolar macrophage. J. biophys. biochem. Cytol. 4, 693—700 (1958).

— Electron microscopic study of the phagocytosis process in lung. J. biophys. biochem. Cytol. 7, 357—366 (1960).

KRAHL, V. E.: Microscopic anatomy of the lungs. Am. Rev. resp. Dis. 80, 24—44 (1959).

KREUZER, F.: Modellversuche zum Problem der Sauerstoffdiffusion in den Lungen. Helv. physiol. pharmacol. Acta 11, Suppl. 9 (1953).

KULENKAMPFF, H.: Bestimmung der inneren Oberfläche einer menschlichen Lunge. Z. Anat. Entwickl.-Gesch. 120, 198—200 (1957).

LAGUESSE, E., and A. D'HARDIVILLIER: Sur la topographie du lobule pulmonaire. Bibl. anat. (Basel) 6, 125 (1898).

LANDAHL, H. D.: On the removal of airborne droplets by the human respiratory tract: I. The lung. Bull. Math. Biophys. 12, 43—56 (1950).

LEBLOND, C. P., and F. D. BERTALANFFY: Reticulin membranes as the framework of alveolar lung tissue in the albino rat. Canad. med. Ass. J. 65, 263—264 (1951).

LENZ, F.: Zur Größenverteilung von Kugelschnitten. Z. wiss. Mikr. 63, 50—56 (1955).

Lewis, B. M., T. H. Lin, F. E. Noe and R. Komisaruk: The measurements of pulmonary capillary blood volume and pulmonary membrane diffusing capacity in normal subjects. J. clin. Invest. 37, 1061—1070 (1958).

Liebow, A. A.: Discussion remark. Amer. Rev. resp. Dis. 80, 41 (1959).

— Pulmonary emphysema with special reference to vascular changes. Amer. Rev. resp. Dis. 80, 67—93 (1959).

Loosli, C. G.: Interalveolar communications in normal and in pathologic mammalian lungs. Arch. Path. 24, 743—776 (1937).

—, and E. L. Potter: Pre- and postnatal development of the respiratory portion of the human lung with special reference to the elastic fibres. Am. Rev. resp. Dis. 80, 5—23 (1959).

Low, F. N.: The pulmonary alveolar epithelium of laboratory mammals and man. Anat. Rec. 117, 241—263 (1953).

— Electron microscopy of rat lung. Anat. Rec. 113, 437—449 (1952).

— Electron microscopy of sectioned lung tissue after varied duration of fixation in buffered osmium tetroxide. Anat. Rec. 120, 827—851 (1954).

— The extracellular portion of the human blood-air barrier and its relation to tissue space. Anat. Rec. 139, 105—123 (1961).

—, and M. M. Sampaio: The pulmonary alveolar epithelium as an entodernal derivative. Anat. Rec. 127, 51—63 (1957).

Macklin, C. C.: Evidences of increase in capacity of pulmonary arteries and veins of dogs, cats and rabbits during inflation of freshly excised lungs. Rev. canad. Biol. 5, 199—232 (1946).

— Alveoli of mammalian lung. Anatomical study with clinical correlation. Proc. Inst. Med. Chic. 18, 78—95 (1950).

— Pulmonary sumps, dust accumulation, alveolar fluid and lymph vessels. Acta anat. (Basel) 23, 1—33 (1955).

— The production of moisture on the pulmonary alveolar wall. Dis. Chest 30, 338—339 (1956).

—, and W. S. Hartroft: The size of pulmonic alveoli based on measurements of their outlines in 25 μ microsections of human and common laboratory animal lungs fixed in the state of expansion. Report to Subcommittee on physiological aspects of chemical warfare, Canada, 1943.

Malpighi, M.: De pulmonibus epistolae II ad Borellum, Bonon, 1661.

Marcus, H.: Lungenstudien V. Morph. Jb. 59, 561—566 (1928).

McNeill, R. S., J. Rankin and R. E. Forster: The diffusing capacity of the pulmonary membrane and the pulmonary capillary blood volume in cardiopulmonary disease. Clin. Sci. 17, 465—482 (1958).

Mead, J.: Mechanical properties of lungs. Physiol. Rev. 41, 281—330 (1961).

Meessen, H.: Die Pathomorphologie der Diffusion und Perfusion. Verh. dtsch. Ges. Path. 44, 98—127 (1960).

Miller, W. S.: The Lung. Springfield, Ill.: Thomas 1937.

Moellendorff, W. v.: Beiträge zum Verständnis der Lungenkonstruktion. Z. Anat. Entwickl.-Gesch. 111. 224—245 (1941).

Moolten, S. E.: A simple apparatus for fixation of the lung in the inflated state. Arch. Path. 27, 116 (1947).

Moore, D. H., and H. Ruska: The fine structure of capillaries and small arteries. J. biophys. biochem. Cytol. 3, 457—462 (1957).

Mueller, A.: Bemerkungen zum Gasaustausch in den Lungen. Helv. physiol. pharmacol. Acta 3, 203—213 (1945).

Neergard, K. v.: Die Retraktionskraft der Lunge, abhängig von der Oberflächenspannung der Alveolen. Z. ges. exp. Med. 66, 373 (1929).

Ogawa, C.: Contributions to the histology of the respiratory spaces of the vertebrate lungs. Amer. J. Anat. 27, 333 (1920).

Orsós, F.: Über das elastische Gerüst der normalen und der emphysematösen Lunge. Beitr. path. Anat. 41, 95—121 (1907).

Otis, A. B., C. B. McKerrow, R. A. Bartlett, J. Mead, M. B. McIlroy, N. J. Selverstone and E. P. Radford jr.: Mechanical factors in distribution of pulmonary ventilation. J. appl. Physiol. 8, 427 (1956).

Pakesch, F., H. Hayek and H. Braunsteiner: Die Struktur der die Lungenkapillaren bedeckenden Epithelhäutchen. Wien. Z. inn. Med. 38, 184—188 (1957).

POLICARD, A.: Le poumon. Paris: Masson 1938.
— A. COLLET and S. PREGERMAIN: Apports de la microscopie électronique à la connaissance histophysiologique de la paroi alvéolaire. J. Physiol. (Paris) 48, 687—690 (1956).
— Sur quelques dispositions inframicroscopiques de l'endothélium des capillaires pulmonaires chez les mammifères. C. R. Acad. Sci. (Paris) 243, 8—11 (1956).
— Etude au microscope électronique des capillaires pulmonaires. Acta Anat. 30, 624—638 (1957).
— Recherches au microscope électronique sur les cellules pariétales alvéolaires du poumon des mammifères. Z. Zellforsch. 50, 561—587 (1959).
—, and P. GALY: Les bronches. Paris: Masson 1945.
PRATT, P. C., and G. A. KLUGH: A technique for the study of ventilatory capacity, compliance, and residual volume of excised lungs and for fixation, drying and serial sectioning in the inflated state. Amer. Rev. resp. Dis. 83, 690—696 (1961).
PROCTOR, D. F., and H. YAMABAYASHI: Degree of lung inflation and capillary filling. Fed. Proc. 20, 107 (1961).
PUMP, K. K.: The circulation of the primary lobule of the lung. Dis. Chest 39, 614—621 (1961).
RADFORD, E. P. JR.: Method for estimating respiratory surface area of mammalian lungs from their physical characteristics. Proc. Soc. exp. Biol. (N. Y.) 87, 58—61 (1954).
— Recent studies of mechanical properties of mammalian lungs. In: Tissue elasticity, J. W. REMINGTON, ed., Amer. Physiol. Soc. 1957, 177—190.
REID, L.: The secondary lobule in the adult lung with special reference to its appearance in bronchograms. Thorax 13, 110—115 (1958).
—, and G. SIMON: The peripheral pattern in the normal bronchogram and its relation to peripheral pulmonary anatomy. Thorax 13, 103—109 (1958).
RICHARDS, D. W.: The aging lung. Bull. N. Y. Acad. Med. 32, 407 (1956).
RILEY, R. L.: Effects of lung inflation on the pulmonary vascular bed. In: Pulmonary Circulation, p. 147—159. New York: Grune and Stratton 1959.
ROBERTSON, J. D.: The ultrastructure of cell membranes and their derivatives. Biochem. Soc. Symp. 16, 3—43 (1959).
— The molecular structure and contact relationship of cell membranes. Progr. Biophys. 10, 343—418 (1960).
ROHRER, F.: Der Strömungswiderstand in den menschlichen Atemwegen und der Einfluß der unregelmäßigen Verzweigung des Bronchialsystems auf den Atmungsverlauf in verschiedenen Lungenbezirken. Pflügers Arch. ges. Physiol. 162, 225—259 (1915).
ROSIWAL, A.: Über geometrische Gesteinsanalysen. Ein einfacher Weg zur ziffermäßigen Feststellung des Quantitätsverhältnisses der Mineralbestandteile gemengter Gesteine. Verh. k. k. Geol. Reichsamt, Wien, 1898, p. 143.
ROSS, B. B.: Influence of bronchial tree structure on ventilation in the dog's lung as inferred from measurements of a plastic cast. J. appl. Physiol. 10, 1—14 (1957).
ROSSIGNOL, 1847: Recherches sur la structure intime du poumon de l'homme et des principaux mammifères. Acad. Med. Belg., Bruxelles, 1847.
ROUGHTON, F. J. W.: The average time spent by the blood in the human lung capillary and its relation to the rates of CO-uptake and elimination in man. Amer. J. Physiol. 143, 621—633 (1945).
SCHEUMANN, K. H.: Zwei Hilfsapparaturen für das petrographische Mikroskop. II. Integrationstisch für das Shandsche Analysenverfahren. Miner. Mitt. N. F. 41, 180—187 (1931).
SCHLIPKOETER, H. W.: Elektronenoptische Untersuchungen ultradünner Lungenschnitte. Dtsch. med. Wschr. 79, 1658—1659 (1954).
SCHUCHARDT, E.: Die Gewebsanalyse mit dem Integrationsokular. Z. wiss. Mikr. 62, 9—13 (1954).
SCHULZ, H.: Elektronenmikroskopische Untersuchungen des experimentellen Lungenödems. Electron Microscopy. Proceed. Stockholm Conf. Sept. 1956, p. 240—243. Uppsala: Almqvist & Wiksell 1957.
— The submicroscopic anatomy and pathology of the lung. Berlin-Göttingen-Heidelberg: Springer 1959.
SCHULZE, F. E.: Beiträge zur Anatomie der Säugetierlungen. S.-B. preuß. Akad. Wiss., Physik.-math. Kl. VI, 225—243 (1906).
SCHWARZ, W.: Elektronenmikroskopische Studien von Fibrillen und Zementsubstanz normaler und silikotischer Lungen. Beitr. Silikose-Forsch. 63, 1—20 (1959).
SHAND, S. J.: A recording micrometer for geometrical rock analysis. J. Geol. 24, 394—404 (1916).

STAMM, W., and K. BUCHER: Lungengewebe und CO_2-Austausch. Acta physiol. pharmacol. neerl. **6**, 261—269 (1957).

STAUB, N. C., and W. F. STOREY: Relation between morphological and physiological events in the lung studied by rapid freezing. J. appl. Physiol. **17**, 381 (1962)

STOREY, W. F., and N. C. STAUB: Ventilation of terminal airunits. J. appl. Physiol. **17**, 391 (1962).

STUTZ, E.: Bronchographische Beiträge zur normalen und pathologischen Physiologie der Lungen. Fortschr. Röntgenstr. **72**, 129—143 (1949).

TENNEY, S. M., and J. E. REMMERS: Comparative quantitative morphology of the mammalian lung: diffusing area. Nature (Lond.) **197**, 54—56 (1963).

THEWS, G.: Die Sauerstoffdiffusion in der Lunge. Ein Verfahren zur Berechnung der Sauerstoffdiffusionszeiten, der Kontaktzeit und des Sauerstoffdiffusionsfaktors. Pflügers Arch. ges. Physiol. **265**, 154—171 (1957).

TOBIN, C. E.: Methods for preparing human lungs expanded and dried by compressed air. Anat. Rec. **3**, 114 (1952).

TOENDURY, G.: Anatomische Vorbemerkungen. In: Handbuch der Inneren Medizin 4, 1—38 (1956).

—, and E. R. WEIBEL: Anatomie der Lungengefäße. Ergebn. ges. Tuberk. u. Lung.-Forsch. **14**, 61—100 (1958).

TOMKEIEFF, S. I.: Linear intercepts, areas and volumes. Nature (Lond.) **155**, 24 (1945) (correction p. 107).

—, and H. CAMPBELL: Calculation of internal surface. Nature (Lond.) **170**, 117 (1952).

THOMPSON, D'ARCY W.: Growth and Form. p. 948. Cambridge: University Press 1942.

VAN ALLEN, C. M., G. E. LINDSKOG and H. G. RICHTER: Gaseous interchange between lung lobules. Yale J. Biol. Med. **2**, 297 (1930).

VEERLOOP, M. C.: The arteriae bronchiales and their anastomoses with the arteriae pulmonales. Acta Anat. (Basel) **5**, 171 (1948).

WATSON, J. H., and V. VALENTINE: Observations on ultrathin sections of dog lung by methods of electron microscopy. Henry Ford Hosp. Bull. **7**, 161—173 (1959).

WEARN, J. T., J. S. BARR and W. J. GERMAN: Behavior of the arterioles and capillaries of the lung. Proc. Soc. exp. Biol. (N. Y.) **24**, 114—115 (1926).

— A. C. ERNSTENE, A. W. BROMER, J. S. BARR, W. J. GERMAN and L. J. ZSCHIESCHE: The normalbehavior of pulmonary blood vessels with observations on the intermittence of the flow in the arterioles and capillaries. Amer. J. Physiol. **109**, 236 (1934).

WEIBEL, E. R.: Die Entstehung der Längsmuskulatur in den Ästen der Arteria bronchialis. Z. Zellforsch. **47**, 440—468 (1958).

— Measurement of alveolo-capillary diffusing surface and of the volume of the capillary bed in normal human lungs. Fed. Proc. **20**, 425 (1961).

— Morphometrische Bestimmung von Zahl, Volumen und Oberfläche der Alveolen und Kapillaren der menschlichen Lunge. Z. Zellforsch. **57**, 648—666 (1962).

— Principles and methods for the morphometric study of the lung and other organs. Laboratory Investigation, **12**, 131—155 (1963).

—, and R. A. VIDONE: Fixation of the lung by formalin steam in a controlled state of air inflation. Amer. Rev. Resp. Dis. **84**, 856—861 (1961).

—, and D. M. GOMEZ: A principle for counting tissue structures on random sections. J. appl. Physiol. **17**, 343—348 (1962).

— Geometry and dimensions of the human airways. Fed. Proc. **21**, 439 (1962).

— Architecture of the human lung. Science **137**, 577—585 (1962).

WENTWORTH, C. K.: An improved recording micrometer for rock analysis. J. Geol. **31**, 228—232 (1923).

WICKSELL, S. D.: On the size distribution of sections of a mixture of spheres. Biometrika **17**, 84 (1925).

— On the size distribution of sections of a mixture of ellipsoids. Biometrika **18**, 151 (1926).

WILLSON, H. G.: The terminales of the human bronchioles. Amer. J. Anat. **30**, 267—295 (1922).

WINKLER, P.: Der Gefäßverlauf in den Lungensegmenten. Schweiz. med. Wschr. **2**, 1341 (1952).

WOODSIDE, G. L., and A. J. DALTON: The ultrastructure of lung tissue from newborn and embryo mice. J. Ultrastruct. Res. **2**, 28—54 (1958).

ZUNTZ, N.: Quoted from BARGMANN 1936.

Subject Index

Morphometry of the Human Lung

by

Ewald R. Weibel, M. D.

Research Associate, Department of Cytology, The Rockefeller Institute, New York, N. Y.
Formerly Research Associate, Department of Medicine,
College of Physicians and Surgeons, Columbia University,
and The Cardiopulmonary Laboratory
of The Chest Service, Bellevue Hospital, New York, N. Y.

With a Foreword

by

André F. Cournand, M. D.

and

Dickinson W. Richards, M. D.

College of Physicians and Surgeons, Columbia University, New York, N. Y.

With 109 Figures

1963

New York

Academic Press Inc., Publishers

Berlin · Göttingen · Heidelberg

Springer-Verlag

Library of Congress Catalog Card Number 63-19191

All rights reserved

No part of this book may be reproduced in any form, by mikrofilm or any other means, without

permission written from the publishers.

© by Springer-Verlag Berlin Heidelberg 1963
Originally published by Springer Verlag OHG Berlin Gottigen Heidelberg in 1963.
Softcover reprint of the hardcover 1st edition 1963

ISBN 978-3-642-87555-7   ISBN 978-3-642-87555-7 (eBook)
DOI 10.1007/978-3-642-87553-3

C'est à vous, cher maître,
Que je voue cet oeuvre.
Vos pensées en étaient les racines –
Par votre génie lui pousseront des fleurs.
J'espère que le tronc et les tiges vous plaisent.

TO MY MASTER AND FRIEND

DOMINGO M. GOMEZ

Foreword

The work presented in this monograph marks a new era, we believe, both in the development of quantitative anatomy of the lung, and in the correlation of anatomy with physiology.

For many years, physiologists interested in the overall functioning of the lung have felt a need for better quantitative descriptions of pulmonary anatomy. As physiologists, we know a good deal about the forces operating to produce pulmonary ventilation, and the quantities that define this function in rest and exercise; and the same for effective distribution of air within the lung — "alveolar" ventilation —, and for the exchange of respiratory gases between air and blood. There have been no correspondingly precise quantitative measurements of the pulmonary structures that serve theese functions.

The great advances in the study of pulmonary anatomy in the past decade have been chiefly in the realm of "fine structure". This has tended to bring together anatomy and biochemistry or physical chemistry, rather than anatomy and physiology. This conjunction has aided, for example, the conception of diffusion as a physicochemical process, but not that of diffusion as a metabolic bodily function.

It was, therefore, a remarkably fortunate circumstance which brought together in our laboratory, about three years ago, Professor DOMINGO GOMEZ and Dr. EWALD R. WEIBEL: Professor GOMEZ a mathematician and biophysicist of distinction and long experience; Dr. WEIBEL a young anatomist trained under Professor GIAN TÖNDURY in Zürich, and with additional research experience with Professor AVERILL LIEBOW at Yale.

As the pages of this monograph will set forth, an entirely new methodology for quantitative description of a complex organ has been developed. To Dr. WEIBEL's own industry and imaginative effort are due new methods of preparation of tissues, new methods of random sampling, and adaptation of methodologies from other fields of science. His collaboration with Professor GOMEZ has led to original mathematical formulations for the translation of the data into the desired geometric forms and a new technique of counting. — Briefly summarized, this work has resulted in reliable information as to the size and number of the basic elements of pulmonary anatomy. More precision as to their number, dimensions, mode of distribution and geometric forms, should eventually facilitate mathematical and physical considerations regarding the function of the lungs. Application of these quantitative methods to the study of pathologic specimens will also provide a most timely renovation of morphologic pathology.

This new era which Drs. WEIBEL and GOMEZ have initiated in the precise quantitative description of the anatomy of the lung, may well be equally applicable to the study of other organs, and may thus eventually establish a broad community of interest between anatomy, pathology, and physiology.

It has been a privilege to have Dr. WEIBEL as one of our research associates for two years, and most gratifying to have had this important work originated and carried out in the Cardio-Pulmonary Laboratory of the Columbia University Division at Bellevue Hospital.

ANDRÉ COURNAND
DICKINSON W. RICHARDS

New York, March 1963

Preface

In discussing with physiologists possible relationships between structure and function of an organ such as the human lung, a morphologist often feels that he speaks a language different in accent and sometimes in vocabulary from that of his colleagues of the other discipline. What appears most interesting — and therefore most important — to him, may be only of secondary significance to the physiologist. In part this may be due to the fact that a morphologist is inclined to marvel at the diversity of structures used to build an integrated organism, while the physiologist will tend to seek general and simple laws underlying complex events, even if this is possible only in the sense of a first and crude approximation.

Intrigued by this apparent incongruity of the goals of pulmonary morphologists and physiologists in particular, I welcomed the prospect of collaborating as a morphologist with a group of outstanding physiologists, and I am most grateful to Dr. ANDRÉ F. COURNAND and Dr. DICKINSON W. RICHARDS for arranging this unusual opportunity for me. While I worked in the Cardiopulmonary Laboratory of Columbia University at Bellevue Hospital they generously supported and stimulated the development of a morphologic study on the lung, whose original goals were necessarily quite obscure. In the course of numerous discussions with their group a long-range program finally evolved: it asked for the search of general principles underlying the structural organization and determining the functional adequacy of the human lung. This appeared to be best achieved through a systematic quantitative analysis of pulmonary structures.

At this point I wish to express my very special gratitude to Dr. DOMINGO M. GOMEZ, to whom this book is dedicated. He actually guided the development of this work onto and along the path it eventually took by first asking the right questions, and then lending a generous helping mind when their answer was to be sought. His role in this work is ubiquitous and cannot be overappreciated. I hope, moreover, that his influence will be a lasting one.

I also wish to acknowledge my sincere gratitude to Dr. GEORGE E. PALADE for allowing me to work and learn in his Department of Cytology at The Rockefeller Institute during this last year and for generously supporting and stimulating the continuation of these studies at the cellular and subcellular levels. I would like to thank Mr. BRUCE W. KNIGHT, Affiliate at The Rockefeller Institute, for his cooperation in the development of necessary measuring principles during this period.

I am thankful to the following granting institutions for supporting this work with generous funds: to the National Institutes of Health for awarding me a Research Fellowship (FFG-47); to the Health Research Council of the City of New York for granting me two Investigatorships (I-126 and I-188); to the New York Heart Association for a Research Grant. Further funds were provided through Training Grant H-2001 of the National Heart Institute, National Institutes of Health, to the Cardiopulmonary Laboratory at Bellevue Hospital.

Finally I wish to express my sincere appreciation to those who enhanced the progress of this work in many ways: to Mrs. BARBARA B. FRANK for her enthusiastic assistance; to Dr. AVERILL A. LIEBOW, Dr. THANE ASCH and Dr. ISRAEL STEINBERG for allowing me to use some of their preparations; to Miss JEAN SEIBERT, Miss CHRISTINE ROSNER and to my wife for helping me with the preparation of the manuscript; to Dr. CHARLES E. LYGHT and Mrs. LOUELLA NIGH for revising the text; and to all my colleagues for numerous suggestions and criticisms.

New York, June 1962   E. R. W.

Table of Contents

From the Department of Cytology of the Rockefeller Institute,
New York, and
from the Department of Medicine, College of Physicians
and Surgeons, Columbia University, New York

Morphometry of the Human Lung

Habilitationsschrift

zur Erhaltung der venia legendi an der Universität Zürich

Vorgelegt von

Dr. Ewald R. Weibel

Anatomisches Institut der Universität Zürich

Zürich 1963